Build the EL Cycle. radioshack.com/DIT

PARTS

- ○ EL Wire Blue (2) RadioShack 2760358
- ○ Power inverter for EL wire/tape/panel (2) RadioShack 2760338
- ○ 12-position European-style mini terminal strip RadioShack 2740680
- ○ Zener diode (4) RadioShack 2760563
- ○ Transformer (2) RadioShack 2731352
- ○ $1/4$" plywood
- ○ Inline skate wheel
- ○ #6 wood screws
- ○ $1/4$" nuts and bolts
- ○ Adhesive tape

- ○ Insulated conduit holders 1" and one wire of each
- ○ Vinyl tubing
- ○ Enclosure RadioShack 2701806
- ○ Heat-shrink tubing RadioShack 2780469
- ○ Speaker wire 2-conductor, search RadioShack.com for "speaker wire"
- ○ Rubber bands
- ○ Zip ties RadioShack 2780472
- ○ Bipolar stepper motor
  http://www.goldmine-elec-products.com/prodinfo.asp?number=G20043
- ○ Bicycle

TOOLS

- ○ Soldering iron RadioShack 6400053
- ○ Solder RadioShack 6400013
- ○ Wire stripper/cutter RadioShack 6400224
- ○ Helping hands RadioShack 6400235
- ○ Heat gun RadioShack 6400212
- ○ Phillips screwdriver
- ○ Adjustable wrench
- ○ Drill and drill bits
- ○ Saw or laser cutter
- ○ Vise

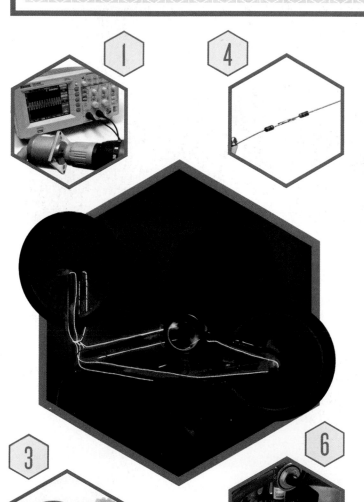

The 12V Zener diodes are connected with the cathodes connected—this will limit the AC waveform at the top and bottom of the cycle. I soldered them, wrapped them in heat-shrink tubing and connected them across the two red (LOW voltage) transformer leads. Clip the leads from two power inverters for EL wire, run them through the lid of the enclosure and connect them to the black (HIGH voltage) side of the terminal blocks. Later the EL wire will be connected to these leads. Each phase of the motor is able to drive a 10' length of EL wire, providing power to a total of 20' of EL wire, which will be wrapped around the bike frame.

Next we have to couple the stepper to the bike frame so that the rear wheel can drive the motor and light the wire. The metal plate that comes with the enclosure is perfect for creating a bracket to hold the motor. To hold the bracket to the bike frame, I used insulated conduit clips. Measure and drill out the holes and mount the conduit clips to the bracket.

Mount the enclosure to the seat tube and the stepper motor-wheel unit to the chain stay. Tape clear vinyl tubing to the bike frame. This will provide added protection for the EL wire and also make it a bit easier to swap out the wire if needed. When you add the wire, make sure you keep it clear of the drivetrain, steering and brakes! Extend the stepper motor's wires using some speaker wire and connect these to the red (LOW voltage) side of the transformers.

radioshack

CONTENTS

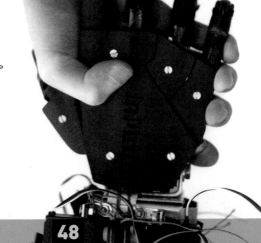

COLUMNS

ON THE COVER:
Nicolas "Bionico" Huchet shows his 3D-printed, Arduino-powered hand. Photography by Miguel Templon

Special Section

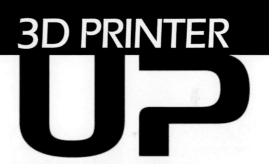

CONTENTS

68

62

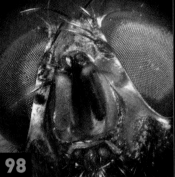

98

80

Vol. 43, January 2015. Make: (ISSN 1556-2336) is published bimonthly by Maker Media, Inc. in the months of January, March, May, July, September, and November. Maker Media is located at 1005 Gravenstein Hwy. North, Sebastopol, CA 95472, 877-306-6253. SUBSCRIPTIONS: Send all subscription requests to Make:, P.O. Box 17046, North Hollywood, CA 91615-9588 or subscribe online at makezine.com/offer or via phone at (866) 289-8847 (U.S. and Canada); all other countries call (818) 487-2037. Subscriptions are available for $34.95 for 1 year (6 issues) in the United States; in Canada: $39.95 USD; all other countries: $49.95 USD. Periodicals Postage Paid at Sebastopol, CA, and at additional mailing offices. POSTMASTER: Send address changes to Make:, P.O. Box 17046, North Hollywood, CA 91615-9588. Canada Post Publications Mail Agreement Number 41129568. CANADA POSTMASTER: Send address changes to: Maker Media, PO Box 456, Niagara Falls, ON L2E 6V2

Make:®

EXECUTIVE CHAIRMAN
Dale Dougherty
dale@makezine.com

CEO
Gregg Brockway
gregg@makezine.com

CFO
Todd Sotkiewicz
todd@makezine.com

> "The urge to miniaturize electronics did not exist before the space program. I mean, our grandparents had radios that were furniture in the living room. Nobody at the time was saying, 'gee, I want to carry that in my pocket'."
> —*Neil deGrasse Tyson*

CREATIVE DIRECTOR
Jason Babler
jbabler@makezine.com

EDITORIAL

EXECUTIVE EDITOR
Mike Senese
mike@makezine.com

COMMUNITY EDITOR
Caleb Kraft
caleb@makermedia.com

MANAGING EDITOR
Cindy Lum

PROJECTS EDITOR
Keith Hammond
khammond@makezine.com

SENIOR EDITOR
Greta Lorge

TECHNICAL EDITOR
David Scheltema

DIGITAL FABRICATION EDITOR
Anna Kaziunas France

EDITOR
Nathan Hurst

EDITORIAL ASSISTANT
Craig Couden

COPY EDITOR
Laurie Barton

PUBLISHER, BOOKS
Brian Jepson

EDITOR, BOOKS
Patrick DiJusto

LABS MANAGER
Marty Marfin

DESIGN, PHOTOGRAPHY & VIDEO

ART DIRECTOR
Juliann Brown

DESIGNER
Jim Burke

PHOTOGRAPHER
Hep Svadja

VIDEO PRODUCER
Tyler Winegarner

VIDEOGRAPHER
Nat Wilson-Heckathorn

WEBSITE

MANAGING DIRECTOR
Alice Hill

DIRECTOR OF ONLINE OPERATIONS
Clair Whitmer

SENIOR WEB DESIGNER
Josh Wright

WEB PRODUCERS
Bill Olson
David Beauchamp

SOFTWARE ENGINEER
Jay Zalowitz

VICE PRESIDENT
Sherry Huss
sherry@makezine.com

SALES & ADVERTISING

SENIOR SALES MANAGER
Katie D. Kunde
katie@makezine.com

SALES MANAGERS
Cecily Benzon
cbenzon@makezine.com

Brigitte Kunde
brigitte@makezine.com

CLIENT SERVICES MANAGERS
Mara Lincoln
Miranda Mota

MARKETING COORDINATOR
Karlee Vincent

COMMERCE

VICE PRESIDENT OF COMMERCE
Kelly Peters

DIRECTOR OF ECOMMERCE
Patrick McCarthy

DIRECTOR OF COMMERCE DESIGN
Riley Wilkinson

RETAIL CHANNEL DIRECTOR
Kirk Matsuo

PRODUCT INNOVATION MANAGER
Michael Castor

ASSOCIATE PRODUCER
Arianna Black

MARKETING

VICE PRESIDENT OF MARKETING
Vickie Welch
vwelch@makezine.com

MARKETING PROGRAMS MANAGER
Suzanne Huston

MARKETING SERVICES COORDINATOR
Johanna Nuding

MARKETING RELATIONS COORDINATOR
Sarah Slagle

DIRECTOR, RETAIL MARKETING & OPERATIONS
Heather Harmon Cochran
heatherh@makezine.com

MAKER FAIRE

PRODUCER
Louise Glasgow

PROGRAM DIRECTOR
Sabrina Merlo

MARKETING & PR
Bridgette Vanderlaan

CUSTOM PROGRAMS

DIRECTOR
Michelle Hlubinka

CUSTOMER SERVICE

CUSTOMER SERVICE REPRESENTATIVE
Kelly Thornton
cs@readerservices.makezine.com

CUSTOMER SERVICE REPRESENTATIVE
Ryan Austin

Manage your account online, including change of address:
makezine.com/account
866-289-8847 toll-free in U.S. and Canada
818-487-2037,
5 a.m.–5 p.m., PST
makezine.com

PUBLISHED BY

MAKER MEDIA, INC.
Dale Dougherty

Copyright © 2015 Maker Media, Inc. All rights reserved. Reproduction without permission is prohibited. Printed in the USA by Schumann Printers, Inc.

CONTRIBUTING EDITORS
William Gurstelle, Nick Normal, Charles Platt, Matt Richardson

CONTRIBUTING WRITERS
John Abella, Alasdair Allan, Leslie Birch, Jordan Bunker, Benton Calhoun, Eric Chu, Jonathan Cook, DC Denison, Stuart Deutsch, Paloma Fautley, Io Flament, Alex Frommeyer, Kate Hartman, Widar Hellwig, Nicolas Huchet, Bob Knetzger, Boris Kourtoukov, Ben Krasnow, Jason Kridner, Bryan Lufkin, Anne Mayoral, Adam McKenty, Eli McKenty, Immanuel McKenty, Isa McKenty, Hans Gerhard Meier, Forrest M. Mims III, Goli Mohammadi, Bo Moore, Stephanie Moyerman, Laura Murray, Dan Royer, Keahi Seymore, Jason Poel Smith, Matt Stultz, Matthew Terndrup, Andrew Terranova, Robert Tu, Lucas Weakley, David Yoon

Comments may be sent to:
editor@makezine.com

Visit us online:
makezine.com

CONTRIBUTING ARTISTS
A.k.a., Jody Culkin, Megan Hellwig, Bob Knetzger, Samantha Lucy, Rob Nance, Charles Platt, Damien Scogin, Peter Strain, Julie West

ONLINE CONTRIBUTORS
Cabe Atwell, Catherine Baxter, Kathy Ceceri, Ian Cole, Jeremy Cook, Marc de Vink, Jimmy DiResta, Lem Fuggit, Travis Good, Agnes Niewiadomski, Luanga Nuwame, Krista Peryer, Haley Pierson-Cox, Isaac Powell, Andrew Salomone, Kyle Scheele, Michael Weinberg, Brian Zweerink

ENGINEERING INTERNS
Enrique DePola, Brian Melani, Nick Parks, Sam Scheiner, Wynter Woods

Follow us on Twitter:
@make @makerfaire @craft @makershed

On Google+:
google.com/+make

On Facebook:
makemagazine

CONTRIBUTORS

What's a piece of wearable technology that you would like to have, but that doesn't exist (yet)?

Kate Hartman
Toronto, Ontario (What to Sense)

I'd like a fitness tracker for rock climbing — something that senses altitude, orientation and tilt of the body, and muscle activity. It could tell me details like how much height I climbed in a given day, how much of it was overhanging, how many falls I took, and what muscles I was using most and least.

Keahi Seymore
San Francisco, California (The Fastest Man on Earth)

I would like some kind of jet boosters you could add to a backpack, body, or boots to allow you to take flight, making huge leaps over miles. Then you could use my Bionic Boots (page 46) as a secondary means of future transportation.

Matthew Terndrup
Los Angeles, California (Six-Way Game Play)

I would like to have a set of wearable gloves that can sense if other accessories like it are near, trading business and contact information with the shake of a hand. The ends of the fingers would also glow, making it perfect for raves on the go.

Peter Strain
Belfast, Northern Ireland (Illustrator, Joseph Gay-Lussac and the Technology of Fireproofing)

I'd like to have the invisibility cloak from Harry Potter — it would really come in handy when trying to avoid any awkward social scenarios and I could secretly pour my own drinks at the pub.

Leslie Birch
Philadelphia, Pennsylvania (4 Fun Flora Projects)

I would be mad-crazy about a stitchable micro-controller with built-in Bluetooth LE, USB, and a JST battery jack. Clothing needs flat surfaces, so stacked shields are cumbersome.

PRINTED WITH SOY INK

DIY Lego Cars and Google Glass, and a More Balanced Makerspace

» I know this is nitpicking, but the "Geek Club" image on page 29 of [Make: Volume 40] has some nits. Assuming the traditional graphic cues for gender hold (big assumption, I know!), there are equal numbers of men and women in the social space. In the workspace, however, men outnumber women six to two, and both of the women are at sewing machines. What's with that? I know our real makerspaces typically have more men than women on the heavy gear, but can't our fantasy makerspaces (there's even a robot serving beer!) be more idealistic?
— *Kim Binsted, Waikoloa, HI*

EXECUTIVE EDITOR MIKE SENESE RESPONDS:
Thanks for the note Kim. That's a good catch — I wish I had noticed it before the magazine published because I agree it'd be great to show women as involved, if not more so, with all aspects in our fantasy makerspace.

Beyond the illustration, though, we do have a lot of female-focused content in the issue. And not by deliberate intent, it just happens that there's a lot of great things happening in makerspaces with women behind them.

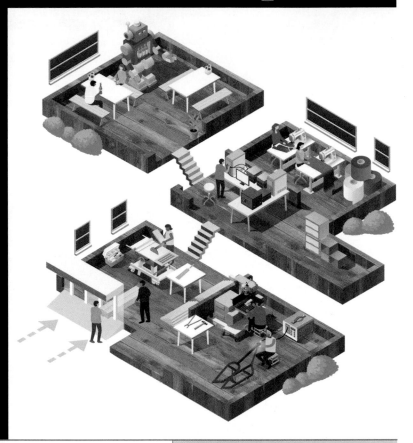

» Hello, my name is Clayton. I am 13 years old. Over the last 2 months I have been working on a DIY Google Glass, inspired by [Make:] Volume 38's section on High Tech DIY. Using an Arduino Esplora and Arduino robot screen I have made glasses that have three voice controls, weather predicting software, and temperature. I mounted the Esplora on the side of glasses I 3D-printed earlier and put a headband on the back to keep it from tipping to one side. Now they are extremely comfortable. In fact, I wear them around my house and tell my parents the temperature just for fun!
— *Clayton, Mississauga, Ontario, Canada*

FROM THE TWITTERVERSE:
» Inspired by the cover of newest issue of @make mag, kiddo made this dc-motor-run #Lego car tonight. Runs great!
— *Rebecca Ahl (@anonyMissBadger)*

MAKE AMENDS:

☐ I KNOW you are aware of this, but on page 86 of Volume 41, Dan Spangler says you get "several thousand pounds per square inch of hold-down force securing your workpiece" when you use a vacuum table on your CNC router. That would be nice, but it ain't possible. I'm sure it was a type or editing error, right?
— *Kevin Hansen Port Townsend, WA*

AUTHOR DAN SPANGLER REPLIES:
Of course you're correct Kevin, what I meant was several thousand pounds of hold-down force securing the entire workpiece on a typical router bed. Thanks for the sharp eye.

☐ In the Tinkering Toys guide in *Make:* Vol. 41, we erroneously named the wood-plank and brackets set on page 39 as Kapla Blocks. These in fact are from Brackitz (brackitz.com), with prices starting at $30. Our apologies. ✿

A Lab Coat that Got Her Dreaming

AS A MAKER, TENAYA HURST IS ALL IN. She's got her own website with a store that sells electronic kits. Her personality shines in a rap video she produced and starred in called "I'm So Maker." She goes by @arduinowoman on Twitter. She teaches a wide range of workshops for adults and children at summer camps and makerspaces. She calls her business Rogue Making, and she is an evangelist for the company that produces Linino, a gig she got through a connection made at her first Maker Faire. Tenaya says that she's always been a maker, yet she recalls the specific day when she discovered electronics. It was March 16, 2013. She was at the Tech Museum of Innovation in San Jose where she worked as a lab instructor and was asked to participate as a facilitator in an Open Make session.

Her colleague, Romie Littrell, had created a conductive lab coat that could be "played" like a musical instrument, just by touching it. He explained that he had put an Arduino in the pocket of the coat and used pieces of conductive fabric as sensors. "When you shook someone's hand, which grounded the circuit, a person could 'play' the pads by touching them," says Tenaya. "It generated all kinds of sounds." The lab coat was a revelation to her and an introduction to the world of wearables. "My mind was completely blown," she says. "You start playing the fabric like a keyboard. I was just so excited." That lab coat set her on a journey that can be an example to young makers everywhere.

After that event, Tenaya sought out people who could help her learn more about Arduino. She attended a workshop organized by Make:SF and learned soldering through a project called Bliplace, a nickel-sized microcontroller with sensors and LEDs that reacts to ambient sound. She started to make her own necklaces and earrings, and she began wearing them in public. They started conversations that otherwise might not have happened. "It gets other people's minds going," says Tenaya. This is what fashion really does, and wearables do it in startling fashion —

> **I WANTED TO HELP PEOPLE REALIZE THAT AFTER THEY DO A FEW EXPERIMENTS, THEY BEGIN TO FIGURE OUT WHAT THEY WANT TO DO.**

attracting people to interact with you.

She loves teaching, confessing that she started teaching electronics to others even before she understood everything she was doing. "I wanted to help people realize that after they do a few experiments, they begin to figure out what they want to do," she says. "You'll start dreaming of your own electronic projects."

Tenaya is more into the electronics than the fashion. "I want to add electronics to garments that already exist," she says, noting that she works with LilyPad. "I like to make things that are interactive."

Her path is one that more girls will follow to discover making for themselves, which can prepare them for careers in science and technology. A new research report from Intel (intel.ly/makehers) coins a wonderful term, "MakeHers," and their survey of tween and teen girls found that one in four has made things using technology in the past year, and seven in 10 would have a desire to learn to create with electronics.

The report notes that girl and women makers are more likely (than men) to discover making through various paths, especially relying on personal connections as resources throughout their making process. The report adds, "girls who make, design, and create things with electronic tools develop stronger interest and skills in computer science and engineering." One of its recommendations is that parents and schools "support and customize making projects based on the identities and interests of participants, whether aesthetic, joyful, or related to helping others." Renee Wittemyer of Intel, who led the research effort, told me:

> "While the report takes a look at the role of girls and women in making, the recommendations support increasing participation and diversity in making at large for everyone. This means inspiring not only girls and women, but underrepresented minorities, boys, and men who may not be excited with 'technology for technology's sake.' Making creates alternate pathways into the fields of computer science and engineering by building on individual interests."

There is a huge need for mentors, both women and men, who help girls and boys develop their dream projects. It's not just about sharing skills, but also, as Tenaya shows us, sharing the passion for making. AnnMarie Thomas, in her book, *Making Makers*, gives this advice to parents: "Whatever it is you love to make, be sure that the children in your lives get to see that passion." It's the stuff that dreams are made of. ●

BY DALE DOUGHERTY, founder and Executive Chairman of Maker Media.

Romie Littrell

Bring the Bids Back Home

Written by Alex Frommeyer

How — and why — your startup should consider onshore manufacturing

IT HAS BECOME COOL TO BUILD THINGS AGAIN. Products like the Pebble watch, MakerBot, and the Nest Thermostat have captured the imagination of the consumer world. New platforms like Arduino and Raspberry Pi make it easier than ever to prototype a hardware idea, and more companies are trying to manufacture those ideas stateside.

Tesla Motors, one of the most provocative companies in the world right now, is a marquee example. One of the most striking features of the company is their vertical integration; its Fremont, California, manufacturing facility houses the production team as well as the engineers and designers. While very common in traditional manufacturing, that practice had become unusual since the late 1970s, when the offshoring movement began to place the designers and the producers of any given widget further (and farther) apart, both geographically and culturally. It has become increasingly common to "design it here, build it there."

Yet, since the Great Recession, perceptions around manufacturing have shifted significantly and today more and more of it is being reshored nationwide.

All of the advancements and interest have produced a nasty side effect, however: Many entrepreneurs without previous manufacturing experience make huge assumptions about the process and its many complexities. From the outside, it seems pretty easy — Wal-Mart and Best Buy are teeming with plastic toys, consumer electronics, and devices for just about anything. Entrepreneurs can be pretty surprised to learn the vast differences between shipping code and shipping a physical good.

So, how can a hardware startup benefit from the resurgence in American manufacturing? When my co-founders and I started Beam Technologies (makers of a connected toothbrush) in early 2012, we had never developed and mass manufactured anything before. We made tons of mistakes, and learned a lot about the process. Perhaps most importantly, we realized that we could learn from Tesla — we too could use vertical integration to our advantage, even with a product far less complex than a car. We decided to bring our entire supply chain within a day's drive of our office, which came with some distinct benefits, both to us and to the suppliers, such as:

- **In-House Quality Control.** Instead of waiting on a shipment to come in from thousands of miles away, our own employees can do quality control in real time. Entrepreneurs can learn directly from industry experts on how adjustments — many of them very low cost — can vastly improve the consistency and quality of the parts coming off the line. Being able to physically meet and compare drawings and prototypes allowed us to resolve simple fitting issues that plagued the first toothbrush.
- **Speed to Market.** Any startup is going to have an irregular and awkward early growth. Changes to parts, molds, and packaging will inevitably need to be made. Having suppliers in the same time zone who can react to your needs quickly can be the difference between a smart pivot to grow faster and a failed business.
- **Build better relationships.** In-person meetings with manufacturers lead to familiarity with, and excitement for, your products. It leads to a congratulatory dinner when the commercial parts roll off the line. And when things go wrong on either end, mutual respect is much more likely to survive. It may also lead to lower upfront costs, delayed payments, payment plans, or other mechanisms to make the financing issues easier. It gives manufacturers a reason to work with you instead of dedicating machine time to a more established player.

Even naïve startups can learn how to build things more efficiently from experienced manufacturing and tooling experts. If the developers and the producers of a product are interacting directly, all of the advantages of quality, communication, and speed should ultimately lead to a better product for your company, and for your customers. ⊘

Samantha Lucy

ALEX FROMMEYER is the co-founder and CEO of Beam Technologies and CEO of Uproar Labs, an Internet of Things R&D firm. An engineer based in Louisville, Alex is obsessed with the Internet of Things, digital health, data science, the maker movement, and startups.

Fully Immersed

Meet the creator of Nomiku, the connected sous vide stick

Written by DC Denison

Hep Svadja

LISA Q. FETTERMAN IS THE CO-FOUNDER AND CEO (CHIEF EATING OFFICER) OF THE HARDWARE STARTUP NOMIKU, which makes "immersion circulators" that allow amateur chefs to affordably experiment with sophisticated sous vide cooking techniques. Fetterman and her two co-founders originally funded their device by raising nearly $600,000 on Kickstarter in 2012. Two years later, they went back to Kickstarter and raised more than $750,000 to fund a new, improved model with wi-fi connectivity and a companion mobile app. Her favorite sous vide dish is miso bass (find the recipe at bit.ly/miso-bass).

Is there something magical about the combination of food and technology that allows you to tap into two very connected communities at the same time?
Yes, that is exactly what is happening. People in the tech community recognize that not a lot of tech is going on in the kitchen. So we are not thinking about incremental innovation — we're not making a phone screen that's just slightly wider. The changes that we're starting to see are super dramatic. Both sides — the food side and the tech side — recognize that and they feel it dearly. When something new comes around in food and tech it definitely makes a splash.

You made the first Nomiku in China. But you've decided to make the next version in the USA. What do you think you'll miss most about China?
In China, the infrastructure is already there. In the U.S., we're going to have to build our own infrastructure.

Prototyping isn't as easy in the US.
No it's not. In Shenzhen, China, where we were, it was like a supermarket for electronics. We lived right next to it, and we'd go in and get whatever we needed. And prototyping, I mean, wow, it was so rapid that it was like I had the thought 10 minutes ago and now I am testing it out on a real prototype.

How did that change when you started prototyping in the U.S.? Are you saving money?
Yes, primarily because of the quality of 3D printers. I can now 3D print a part for around $300, and put everything together into a prototype for like $600 to $800. That's for a fully functional prototype, versus the $6,000 that it cost prototyping with our contract manufacturer in China.

Has working on the Nomiku project caused you to have new respect for everyday cooking gadgets?
Actually, I have always been amazed by things like a $6 toaster. How can it be so cheap, and still work as advertised? Just achieving that is so hard. I am constantly amazed by the Vitamix blender, which is made in the United States. I think they are a bastion of hope for everybody who wants to make a high-tech kitchen appliance. It is so durable and does exactly what it says it will do — nothing more, nothing less. That's what makes people happy. But getting there is so hard.

Your new model allows users to connect to it over wi-fi via a mobile device. Is that just an Internet of Things gimmick?
Not at all. What I like about the Internet of Things is that it connects everyone. Community is what makes food awesome. We always wanted to bring that with our appliance and now we literally can. We're making it easier to share recipes with top chefs, great scientists, and your next-door neighbor.

Cooking and the internet are so natural together, and the Internet of Things is a way to make that happen — it's actually connecting you to other cooks. I think everybody will benefit if every single piece of technology has this kind of community approach. ◐

DC DENISON is the editor of the Maker Pro Newsletter, which covers the intersection of makers and business, and the former technology editor of The Boston Globe.

The Worst Maker in the Room

Too much planning leaves too little time for making. Written by Hans Gerhard Meier

HANS GERHARD MEIER can be found in the workshop making mistakes when he is not teaching at the Oslo School of Architecture and Design in Norway.

I'M GENETICALLY PREDISPOSED TO DO THINGS TWICE. My dad would eagerly make stuff, only to have my mom tell him to do it over again, nicer and better. I could make a long list of stuff I've made that failed, including some that scared my wife and scarred me.

Ideas for new projects come at any time of day or night; my sketchbook keeps me on track. Jotting down possible solutions for ideas is the start of any project. I make it a rule not to get too detailed — I'd rather be testing and making as soon as possible. It's better to act on that initial ignition before new projects pop up. Going from sketch to dirty hands needs to happen fast. This process leads to some obvious fails, but those are part of the process. A too-detailed idea on the drawing board will result in hurdles before I even begin, so I'd rather try, fail, and retry. The mistakes never stop me; they fuel a wish to overcome the problem and learn firsthand. I will sometimes be advised to drop my proposed solution, but like a child I will stubbornly try anyway. I love figuring out what went wrong and doing it all over again. When I finally get it right, I've become a better maker, ready for the next level.

Approaching new projects with a naive attitude gives me a chance to refine my task as I go. Learning by doing is a great mantra; this involves failing to succeed. My wife went away one weekend, and I set out to make a foldable ladder for our attic. After a quick sketch to figure out what materials I needed, I started building. Hinges and locks were essential parts of the plan. After hours putting it together, and attaching the ladder

> ## "Going from sketch to dirty hands needs to happen fast."

to the ceiling, I got ready for the one small step. Luckily, I was only halfway up when the thing collapsed and I fell to the floor. My sore back and wounded maker pride faded by the day's end and what was left was the joy of failing, learning firsthand the impact 165 pounds have on a foldable ladder. We have a ladder now, though it's not foldable. It's not as impressive as my first idea, but I make it up and down from the attic alive.

A mantra I continuously repeat is quantity equals quality. The more I make, the bigger the chance that I make something great. I have boxes filled with sketchbooks, a testament to this philosophy. They are filled with stupid, crazy, and useless ideas, but once in a while combining two of them makes gold.

I love sharing my mistakes. I brag when they work, but also when they "work." The mishaps and redo's open up a fruitful discussion. People will help and give advice, telling stories of their own misses and fails. A great mistake means you're aiming high, and by sharing, you will most probably succeed.

Over the years I've wound up with many misfit projects, but I love them all. I've come up with a name for these hard to describe projects, partly to justify to my wife the many hours spent on making an electrical ear-cleaner. When friends and family ask why I made something, my best reply is "because." So I made up a new word (in Norwegian): *Hvorfordi*, which translates into two words, why (*hvorfor*) and because (*fordi*). It doesn't matter what you make — or whether you're any good at it — as long as you make it. It's making for making's own sake. When I failed a typography course at San Francisco's Academy of Art college, the teacher who failed me told me something I still strive to live by: "Think it, make it, and move on." Some day I'll make that a tattoo. ◉

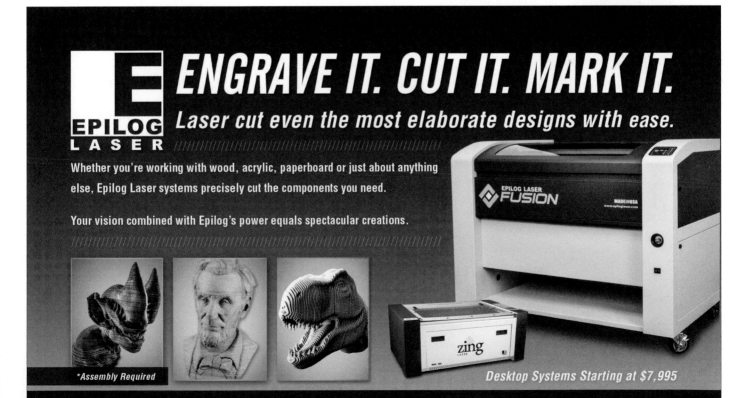

MADE ON EARTH

The world of backyard technology

Know a project that would be perfect for Made on Earth?
Email us: *editor@makezine.com*

BRICK MENAGERIE

SEANKENNEY.COM

You can jazz up your garden in lots of ways: A bench; a tasteful birdbath; a storm-resistant lawn gnome.

Or a bison that's the size of an NBA player and is made of 45,143 Legos.

Artist **Sean Kenney**'s solo exhibition, *Nature Connects*, is touring botanical gardens and zoos all over the U.S. It features huge foxes, frogs, butterflies, and bees constructed entirely of the Danish building blocks.

"Unlike traditional sculpture, you can't just carve out a shape or add to a surface," Kenney explains. "You have to think ahead as you're building upwards linearly." The artist wanted to show relationships within nature. (He also just loves Legos — he's written seven children's books about them.)

The plastic fauna has been on tour, traveling from Nebraska to Florida, since the spring of 2012, and will migrate across America until 2019, flocking to cities like San Francisco, Atlanta, Salt Lake City, and San Antonio.

— *Bryan Lufkin*

Sean Kenney

MAGICAL, MUSICAL POLYHEDRON

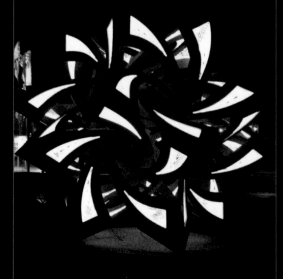

When mathematician George Hart created his 11-inch-diameter dodecahedron sculpture titled *Frabjous*, he had no way of knowing it would inspire an Austin, Texas-based group of makers, led by artist **Ilya Pieper**, to create a 10-foot-diameter interactive, musical version. Named *The Cathedral of Celestial Mathgic*, it's a celebration of the inherent beauty of math. The supersized polyhedron is at the center of a 50-foot-wide pentagonal structure with four arches shaped like lotus petals.

Participants are encouraged to experiment and play with the sculpture, engaging sensors to control lights and sounds via an Arduino. It's been installed at Burning Man, Art Outside, and SXSW, and Pieper says she's even witnessed a large group of people being orchestrated to play all the points simultaneously.

The Cathedral of Celestial Mathgic took a core team of eight makers (plus friends) roughly nine months to go from concept to creation, including grant writing and fundraising. "It's been pretty incredible to witness how much effect this piece has had on so many people," says Pieper. "I don't think I ever could have imagined."
— *Goli Mohammadi*

Aaron Geiser

BUSINESS REPLY MAIL
FIRST-CLASS MAIL   PERMIT NO. 865   NORTH HOLLYWOOD, CA

POSTAGE WILL BE PAID BY ADDRESSEE

Make:

PO BOX 17046
NORTH HOLLYWOOD CA 91615-9186

The future is destined to bring us technology composed of fungus, according to **Phil Ross**, a mushroom artist and engineer of sorts. "A living computer sounds so crazy but its also very reachable," he says.

Ross has been experimenting with mushrooms since the '90s, when he became interested in their medicinal properties and health benefits. Once he found they were infinitely renewable and understood the materials could conform to the shapes and conditions of the environment, the artist began dreaming up more complex forms.

Mushrooms, Ross says, are a "self-extinguishing organic material" that can be used to create habitats for humans on Earth or even in space. He is currently working on a prototype, envisioning a demonstration building on the San Francisco waterfront to show how mushrooms can be used as building materials.

"When we're done with it, we can just push [the building] into the bay," Ross says. "Mushrooms change the politics and aesthetics of pollution."
— *Laura Rena Murray*

Phil Ross

SIX-WAY GAME PLAY

BRENTBUSHNELL.COM/PROJECTS/PARTY-TABLE

Everybody gather round the Hexacade, a six-person, custom gaming table created by **Brent Bushnell** and **Eric Gradman**, founders of Two Bit Circus, an "entertainment engineering" company. The console was designed as a way for people at parties to meet each other. "Most people come to events in groups of two's and four's," says Bushnell. "Six was a special number because six guarantees meeting someone new."

The outer frame was originally made of wood, but was later updated by repurposing an aluminum equipment case to hold the circuitry inside. An Intel NUC mini computer runs the programs, which are coded in Python. The games include ping-pong, marble racing, and a light-cycle experience (similar to Tron). The black vinyl decal around the controls was etched with a laser cutter, and the outside body was made with a CNC router. Clear acrylic sits on top as a protective layer, and the whole setup comes is on wheels — Two Bit Circus made several units, which they keep in a warehouse and roll out to events. — *Matthew Terndrup*

Two Bit Circus

HEART OF STEEL

Babette Daniels

MAKEZINE.COM/
STEAMPUNK-ROBOT

Thomas Willeford, of Brute Force Studios, is a giant among makers in the steampunk community. When **Jeff Mach**, organizer of the Steampunk World's Fair, wanted a giant robot, he enlisted Willeford to create it.

Over 15 feet high and more than 1,000 pounds, the robot is constructed of ¾-inch plywood sheathed in Masonite. It is finished with furniture tacks and painted to look like riveted metal. Though unmanned, the torso is big enough to fit a person inside, and the shoulders, elbows, and wrists are poseable. It took 45 12-hour days to build, with 17 people assisting at one time or another.

"I intended the robot to be a symbol of how creative steampunk can be," says Mach. He also wants to support cancer awareness in the steampunk community. A plaque on the robot is dedicated to "Hearts of Steel," listing, among others, CJ Henderson, a writer and editor of steampunk literature who recently lost his battle with cancer.

— *Andrew Terranova*

WIRES& THREAD

ELECTRONICS SHRINK. POWER REQUIREMENTS LESSEN.
Capabilities compound. Processing speeds skyrocket. Interfacing options multiply. Today's microcontrollers and microsized computers greatly surpass those of even just a year ago, making almost any project feasible. And some of the hottest projects right now come from the world of wearables.

What is a wearable, exactly? Some call the space "wearable electronics," others "wearable computing." We see it as something far broader than that. Wearable technology can be advanced electronic sensors and displays combined with everyday apparel, like Io Flament's brainwave-sensing beanie (see page 60) and Jonathan Cook's Open-Source Smartwatch (page 54), but it also includes the futuristic under-the-skin projects of a few daring body augmenters (page 53), as well as the mechanical technology in Keahi Seymour's Bionic Boots (page 44).

In this section we will guide you through these examples and more. But first we'll explain all the latest trends in electronics prototyping boards — including advanced wireless communication protocols — and which boards to keep an eye on. All of them smaller and more powerful than ever. ◗

A SMORGAS-BOARD

Wearables and IoT are driving diversity in the microcontroller market

WRITTEN BY ALASDAIR ALLAN

IF YOU'RE OLD ENOUGH TO REMEMBER THE '80S AND THE HUGE RANGE OF PERSONAL COMPUTERS that sprung up in the early years of the industry — each from a different manufacturer and based around a different CPU — then the current state of the microcontroller board market may seem familiar. In the last year and a half, we've seen an explosion of new boards, and there doesn't appear to be any reason to expect the trend to slow down in the near future. But the microcontroller market isn't the personal computer market. The forces driving change are very different, and because of that, it's unlikely that we'll trade the current diversity for a monoculture. Instead, something much more interesting is going to happen.

THE THIRD ESTATE

Most of the new boards being released will disappear — almost without a trace — within a few months. The ones that do hang around tend to be those that form communities around them. Right now the two largest of these communities belong to the Arduino and Raspberry Pi. Even as other boards are doing interesting things and have sizable presences in some markets, those are the two names that tend to dominate the conversation around microcontrollers and single board computers.

Building communities from scratch is hard. So for a new board to become commonplace it typically has to either win over an existing community or co-opt a community that exists but isn't tied to a specific board. A growing number of new board manufacturers are taking the latter path, appealing to a community not traditionally associated with makers: web developers.

Perhaps surprisingly, there's a history of hardware hacking inside the Node.js and Javascript communities. But with the arrival of the Tessel and the Espruino, both of which run Javascript right on the board itself, developers can now hack hardware in their native language.

However, unlike the communities that have grown around the Arduino and the Raspberry Pi, the thing that holds this community together is the language, not the board. It stands to reason then that this third community might not settle on a single board. There's room for more than one player and possibly more than one way to do Javascript.

Hep Svadja

ALASDAIR ALLAN
is a scientist, author, hacker and tinkerer, who at the moment is spending a lot of his time thinking about the Internet of Things. He has mesh-networked the Moscone Center; caused a U.S. Senate hearing, and contributed to the detection of what was — at the time — the most distant object yet discovered.

ESPRUINO

BEAGLEBONE BLACK

**SPARKCORE
(WITH CC3000)**

**SPARK.IO
PHOTON**

LIGHT BLUE BEAN

The BeagleBone Black, which some see as the main competitor to the Raspberry Pi, is putting a lot more effort into publicizing the inclusion of the Cloud9 development environment and its own BoneScript — a Node.js library specifically optimized for the Beagle family and featuring familiar Arduino function calls. Similarly, other Linux-based boards such as the WeIO are now supporting Javascript-based development environments out of the box.

WIRELESS EVERYWHERE

For a long while, getting networking of any kind to work with the Arduino was pretty difficult; getting wireless networking to work was more difficult still. In my view, this is one of the factors that contributed to the maker popularity of the Raspberry Pi, as well as Digi's XBee radios.

With a new generation of wireless radios arriving in the market — notably Texas Instruments' CC3000 — things have changed radically. It's just not that hard anymore. Almost every new board comes with a built-in radio of some kind, with Bluetooth LE or wi-fi being the most popular.

Good examples of this include the Light Blue Bean, the only Arduino board I know of where uploading the sketch is done over Bluetooth LE, and the new Spark.io Photon board, which has onboard wi-fi in a package the size of a postage stamp. Priced under $20, it's almost a poster child for the next generation of network-enabled microcontroller boards.

WEARABLES AND THE INTERNET OF THINGS

The reason the microcontroller market probably isn't going to evolve into a monoculture is that boards aren't general-purpose computers. Microcontrollers are used to control a variety of things, and that means that there isn't a single-use case. There will almost inevitably continue to be a lot of diversity in both architecture and form factor.

One thing that's driving that diversity at the moment is the rise of both wearable electronics and the Internet of Things. Both of these trends are having a huge influence on microcontroller designs, with boards tending to be smaller, more power-efficient, and have built in radios — essential for wearable-use cases where you can't plug your board into a wall socket. You can see this clearly in the current generation of boards appearing on Kickstarter and other crowdfunding platforms, including the

EDISON

MICROVIEW

CLOUDBIT

MetaWear, the MicroView, and, of course, the Light Blue Bean again.

A good example of taking this to an extreme is the ESP8266 wi-fi serial transceiver module. At just $5, this tiny module was originally intended to allow you to connect your project to the internet cheaply and simply. But the addition of a community-built GCC SDK makes it a microcontroller board in its own right — and a very affordable platform for the Internet of Things.

MAKER TO MAKER PRO

Despite what you might have read elsewhere, the maker movement didn't invent the microcontroller. There's a vast — and previously unreachable — enterprise market out there, a market that's happy with a bag of surface-mount chips and an indecipherable 400-page manual. One of the interesting developments within the last year or so is that this market has started to pay attention to things going on in the maker world. And the maker world has started to take notice of it.

The evidence of this comes from two directions and converges in the middle. From the maker world we have the Raspberry Pi Compute Module — a board intended to allow you to embed the brains of the Pi into a commercial product. While from the industrial world we have Intel's Edison board,

which pretty much has the same mission (see page 26).

We're already seeing commercial products such as the OTTO camera, which integrates the Compute Module, spun out from the maker community. And no doubt we'll see the same from the Edison module once it's more widely available.

NO PROGRAMMING REQUIRED

At the opposite end of the spectrum are products such as littleBits that require virtually no programming. Initially seen by many as a toy, the system is now being taken far more seriously. Several larger companies have even used it to take ideas from mock-ups to actual products.

The introduction of the Arduino bit and, more significantly, the cloudBit — which adds the ability to snap an internet connection onto just about anything and automate it — makes the system simple to get to grips with and infinitely flexible. The arrival of the bitLab, a marketplace for user-generated bits, opens the system to both makers wanting to tinker and pro makers hoping to add unique abilities.

Similar systems such as the Kickstarter-funded BLE-connected range from SAM labs and the Wunder Bar Internet of Things starter kit from Relayr are now starting to

appear. However you do have to wonder whether they'll take off. With several years' head start, littleBits has had a big advantage building its community.

ARDUINO AND RASPI, THE NEXT GENERATION

With large communities heavily invested in the current form factors and software stacks, both Arduino and the Raspberry Pi Foundation have less room to be radical about their next steps. However it's not just more of the same from the two big players in the board market.

Due to arrive on the market soon, the new Arduino Zero is built around an ARM processor. It's obvious that this board is intended, with time, to replace both the Uno and the Leonardo and move the company to a 32-bit platform. Beyond that, the Arduino at Heart program will allow them to control the explosion of next generation Arduino compatibles like the Apollo board.

The Raspberry Pi Foundation also has released new model A+ (see page 26) and B+ boards, replacing the original model A and B respectively. In the two years since the launch of the original Pi, there have been a lot of complaints about how the board was put together, although none of these problems seemed to hinder the massive success of the board. Nevertheless, the new model

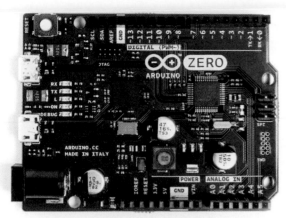

ARDUINO ZERO

RASPBERRY PI B+

APOLLO

is aimed at putting these criticisms to rest, once and for all.

Beyond the two new models, I think we could well expect further — perhaps more specialized — boards from the Raspberry Pi Foundation. However I see the introduction of their HAT standard for expansion boards as locking them into the current form factor for some time to come.

WHERE NOW?

I think the next six months to a year is going to see the board market evolve along similar lines to the previous 12 months. The proliferation of board designs will continue and, if anything, we should expect more new boards, not fewer. However most of these new boards are going to disappear just as rapidly as they arrived, and beyond this we might see a period of consolidation in the marketplace.

In the near term, the trend towards smaller boards and onboard radios will continue. The Espruino, for instance, will soon come in Pico. Although I do wonder whether, as with the Arduino before it, we'll see some sort of default standardization in the footprint of these smaller boards. And perhaps from there a fourth community will form around a board in the wearables space. ◗

INTEL EDISON:
RAPID PROTOTYPE

The surprising maker backstory of the tiny, powerful new computer

WRITTEN BY DAVID SCHELTEMA

BRIAN KRZANICH STOOD IN FRONT OF A PACKED AUDIENCE AT THE 2014 CONSUMER ELECTRONICS SHOWCASE in Las Vegas, backed by a sea of oversized blue display screens. The Intel CEO had just finished presenting a smartwatch and headset — new wearable projects that the company was collaborating on — when he reached into his pocket and pulled out what appeared to be an SD card.

The CES cameras steadied focus on the tiny board between his fingers as Krzanich explained that what he was holding was actually a full Pentium-class PC. The module was equipped with wi-fi, Bluetooth, and more, he continued, all designed to help people create powerful new tech products quickly. He called it the Edison.

The crowd loved it, and the tiny computer became one of the big stories of the January event. Its reveal had all the slickness of any prominent consumer-device product launch. But beneath its tidy package the Edison has the surprising imprint of the maker movement.

The story begins in Portland, Oregon, in January 2013 at the biannual Intel TechFest, an internal networking conference for senior engineers. A team from Intel labs in China, led by Sun Chan, presented a stamp-sized microcontroller called the PIA (short for Pervasive Intel Architecture) — a board with huge potential for makers.

Michael McCool, a software architect and principal Intel engineer based in Tokyo, took interest in the small chip with powerful processors. McCool, who often develops and demos projects at Maker Faires, fostered a relationship with Chan and his team, and began to consult on the project, providing technical expertise along with insights for mechanical design.

The now multinational PIA team continued to work on the project, refining technical specifications, and soon after found a new advocate in Intel CEO Krzanich. "One of our teams from Intel labs in China knew I was a maker and interested in the maker movement, so they brought Edison to me for a review," Krzanich explains. "When I saw it I knew where it would lead us," he continues, hinting at multiple development kits and online community resources.

With Krzanich's go-ahead came a new deadline: The PIA team had to have a prototype ready to be unveiled at CES 2014, just three months away.

One particularly challenging aspect of transitioning the board from prototype to product, explains McCool, was that the team wanted to retain the original intention of an SD card form factor. "The problem is they couldn't find any cases that could fit around it because the tolerances were so tight. The off-the-shelf cards wouldn't fit."

In true maker fashion, McCool tackled the problem by designing a cast-resin mold and embedding the card into epoxy, utilizing a workspace that was a far cry from the gleaming-white clean rooms shown in Intel commercials. "The masters for these molds were milled in the Tsukuba, Japan Intel site, but the actual encapsulation and pouring were done on [my] kitchen table," he explains with a laugh.

But even with custom molds, obstacles remained. For example, the silicone sealant he used held the PCBs in place by their exposed contacts during pouring. "It was a real challenge to get the resin to flow into the tiny space around the prototype PCB." After much experimentation, he found that first heating the resin in a warm-water bath reduced the viscosity significantly.

As the team scrambled to get ready for the CES 2014 reveal, other groups at Intel worked to take Edison from a lab initiative to a consumer good, looking for companies who could integrate the board into a new product for the launch.

Ed Ross, senior director of inventor platforms for the New Devices Group, found just the team for the task in Rest Devices, a Boston-based startup with a specialty in rapid prototyping. The firm jumped into the assignment, met the one-month deadline, and joined Krzanich onstage at CES to demo a baby-monitoring onesie called Mimo which uses Edison to handle sensor and communication I/O.

After the CES debut, the team continued to refine and augment Edison's capabilities, swapping the original Quark processors for the newer, more powerful Atom cores and adding RF shielding absent in the epoxy-enclosed prototypes. Ultimately, these upgrades required a change in size and shape, but the diminutive computer still impresses at just 1.4"×1". Intel officially released Edison to the public in September 2014, just 11 months after Krzanich had given the project the green light. ✏

Prototype Intel Edison in an epoxy enclosure

PIA team in December 2013

Preparing an early version of the Edison for resin encapsulation using silicone molds. The method was used in lieu of standard cases. Polishing the resin made it possible to see the internals.

Rest Devices' Edison-powered Mimo Smart Baby Monitor, launched with Intel at CES 2014

PHOTOGRAPHIC MEMORY

WRITTEN BY STEPHANIE MOYERMAN

STEPHANIE MOYERMAN is a research scientist with the Smart Device Innovation Team in Intel's New Devices Group.

Hep Svadja

Teach Intel's Edison how to spot faces using OpenCV

COMPUTER VISION IS A PROCESSOR-DEMANDING TASK, but thanks to a dual-core Atom processor, the Intel Edison handles it with ease. The Edison ships with a highly custom Linux image, but you'll only need to add a few software packages and custom code to get OpenCV — a wildly popular approach to computer vision — operational and recognizing human faces in photos.

1. FLASH THE EDISON WITH THE LATEST FIRMWARE
Follow the flashing instructions on the Intel documentation page at makezine.com/go/flashing-edison to update your Edison with the latest image.

Then run the Edison configuration script:

```
configure_edison --setup
```

And follow the setup prompts to configure a hostname and root password and to set up wi-fi access.

2. SSH INTO THE EDISON
On Windows, download and install Putty, an SSH client. Then point Putty to your Edison.

On OSX or Linux, open a terminal and type:

```
ssh root@edison.local
```

NOTE: If you changed the hostname, replace `edison` in this address with the new name you created.

3. INSTALL THE LATEST IOT DEVELOPER KIT LIBRARIES
Type in the following commands, and note this is actually one long line with spaces between `intel-iotdk` and the URL, and on both sides of the > character:

```
echo "src intel-iotdk http://iotdk.intel.com/repos/1.1/
intelgalactic" > /etc/opkg/intel-iotdk.conf
```

Update the package repository, then upgrade all the packages:

```
opkg update
opkg upgrade
```

4. ADD AN UNOFFICIAL PACKAGE REPOSITORY
Access to every package is not available without adding repository locations to the *opkg/base-feeds.conf* file. By doing this, you'll add an enormous number of compiled applications, saving you the hassle of compiling from source.

NOTE: Unofficial repositories are quite common across most Linux distributions.

Add the following lines to *base-feeds.conf*:

```
echo "src/gz all http://repo.opkg.net/edison/repo/all
src/gz edison http://repo.opkg.net/edison/repo/edison
src/gz core2-32 http://repo.opkg.net/edison/repo/core2-
32" >> /etc/opkg/base-feeds.conf
```

Update the repository index again, since you just added new

```
package locations:
  opkg update
```

Next, install NumPy, OpenCV, and OpenCV-Python.

```
opkg install python-numpy opencv python-opencv nano
```

That's it! All the necessary packages are installed. Time to start hacking code!

NOTE: Installing the basic text editor **nano** is not necessary, but is suggested unless you're comfortable using **vi.**

5. PROGRAMMING WITH PYTHON AND OPENCV
Launch nano and specify a filename to use. Then import the 3 required Python libraries:

```
nano ~/FaceDetection.py
  import numpy
  import cv2
  import urllib
```

Download and place our sample photo (seen here at right) in the Edison's web server directory with the new filename, *in.jpg.*

```
print("Downloading Images and Necessary Files")
urllib.urlretrieve(http://cdn.makezine.com/make/43/
Intel_CES_Team.png, '/usr/lib/edison_config_tools/
public/in.jpg')
```

Next, download the XML file that defines the parameters for the OpenCV facial-recognition algorithm. This file is also saved to the public directory of the Edison's web server as *haarcascade_frontal face_alt.xml.*

```
urllib.urlretrieve('https://raw.githubusercontent.com/
Itseez/opencv/master/data/haarcascades/haarcascade_
frontalface_alt.xml', '/usr/lib/edison_config_tools/
public/haarcascade_frontalface_alt.xml')
```

Import the photo using OpenCV and convert it to grayscale for use in the facial-recognition processs:

```
img = cv2.imread('/usr/lib/edison_config_tools/
public/in.jpg')
gray = cv2.cvtColor(img,cv2.COLOR_BGR2GRAY)
```

Using the OpenCV libraries, create the facial-recognition algorithm and process the grayscale image:

```
faceCascade =
  cv2.CascadeClassifier('haarcascade_frontalface_
  alt.xml')
faces =
  faceCascade.detectMultiScale(gray,scaleFactor=1.1,minN
  eighbors=5,
  minSize=(30, 30), flags = cv2.cv.CV_HAAR_SCALE_IMAGE)
```

The **faces** variable now contains an array of rectangular coordinates that surround each face that OpenCV found in the image. These coordinates are then used to draw a box around each face in

The Intel demo team at CES posing with some of the awards the Edison won

the original color image, which you'll save as a new file:

```
for (x,y,w,h) in faces:
    cv2.rectangle(img,(x,y),(x+w,y+h),(255,0,0),2)
  cv2.imwrite('in_facefound.png',img)
```

Finally, save the text file by pressing Ctrl-X on your keyboard. When prompted to save the file, type Y and Enter.

6. WEB PAGE SETUP
Download a simple HTML file which will display the pre- and post-processed images on the Edison's onboard web server.

```
wget http://cdn.makezine.com/make/43/OpenCV.html
```

Change directories to the web server's public directory:

```
cd /usr/lib/edison_config_tools/public
```

7. VIEWING THE IMAGES
Now head on over to http://edison.local/OpenCV.html to view the Before and After images, with a box around each detected face!

GOING FURTHER
Now that OpenCV and Python are configured on your Edison, be sure to see the official documentation for great example code and ideas at makezine.com/go/opencv-python-tutorials. OpenCV can detect all kinds of shapes, analyze video, and much more. ●

DIY
CONDUCTIVE INK

TIME REQUIRED:
2 DAYS
COST:
$45–$130

MATERIALS

- » **Silver acetate (99%), 1g**
- » **Ammonium hydroxide (28%-30%), 3.0mL**
- » **Formic acid (88% or higher), 0.5 mL**
- » **Wood, ½"×3"×3"** or larger
- » **Bolt, 2"**

TOOLS

- » **Band saw**
- » **Bench vise**
- » **Hacksaw**
- » **Hot glue gun**
- » **Test tube with stopper**
- » **Small glass vial**
- » **Beaker (2)**
- » **Dispensing syringe, 100mL (3)**
- » **Syringe filter, 0.2µm**
- » **Weight boat**
- » **Digital scale**
- » **Drill**
- » **Neoprene Gloves**
- » **Safety goggles** rated for chemical splashes

JORDAN BUNKER
is a polymathic jack-of-all-trades who enjoys manipulating ideas, atoms, and bits. When he's not braving the daylight, he can be found in his basement workshop in Seattle.
www.hierotechnics.com

Illustrated by James Burke

Mix up a homemade batch using some basic chemistry skills

WRITTEN AND PHOTOGRAPHED BY JORDAN BUNKER

THANKS TO RECENT SCIENTIFIC ADVANCES, YOU CAN BUY CONDUCTIVE INKS IN THE FORM OF PENS, PAINTS, AND EVEN PRINTER CARTRIDGES, but have you ever wondered if you could make your own?

You can, and following a simple process developed by the University of Illinois Urbana-Champaign Materials Research Laboratory, it's actually quite easy to produce the conductive ink at home.

The following steps have been adapted from the UIUC paper titled "Reactive Silver Inks for Patterning High-Conductivity Features at Mild Temperatures," and have been simplified for the amateur chemist.

> **CAUTION:** The chemicals used are strong-smelling, corrosive, and can stain skin and clothes. You must wear chemical-splash safety goggles, neoprene gloves, a long-sleeve shirt, pants, and close-toed shoes. Due to the fumes created, this project must be done outside or in a fume hood.

PREPARATION

Clean all glassware and tools and lay them out on your work surface. It is important to read through all of the steps, and make sure that you understand them thoroughly before beginning .

1. MAKE A VORTEX MIXER

Rather than using an expensive laboratory vortex mixer, you can make your own using a 2" bolt and a circular piece of wood.

Cut a circle approximately 2½" in diameter from a piece of ½" thick wood. Drill a centered hole large enough to fit the shaft of the 2" bolt. Drill a second ½" hole halfway into the wood, slightly off-center and overlapping the centered hole (Figure **A**).

Place the 2" bolt in a bench vise, and then use a hacksaw to remove the bolt head. Insert the bolt shaft into the center hole until it is flush with the bottom of the ½" hole, and then secure it with hot glue.

2. MAKE THE INK

Pour roughly 3mL of ammonium hydroxide into a glass beaker. Using a dispensing syringe, draw exactly 2.5mL out of the beaker and deposit it into the test tube (Figure **B**).

Place the weight boat on a digital scale and tare the scale. Measure out exactly 1g of silver acetate powder (Figure **C**) and pour it into the test tube.

Insert the bolt of your homemade vortex mixer into the electric drill. Holding the top of the test tube firmly, place the base of the test tube against the hole in the vortex mixer (Figure **D**). Slowly increase the speed of the drill until a vortex appears in the test tube, mix for 15 seconds, then set the test tube aside.

Pour approximately 0.5mL of formic acid into a second glass beaker. Using a new dispensing syringe, draw exactly 0.2mL of formic acid out of the beaker (Figure **E**).

Drip one drop of formic acid into the mixed solution in the test tube (Figure **F**), then vortex-mix using the same method from step 2. Repeat this process until all 0.2mL of formic acid has been mixed.

After mixing, the solution will be a gray or black color. Place a stopper in the top of the test tube, and set the test tube aside to react for at least 12 hours (Figure **G**, following page).

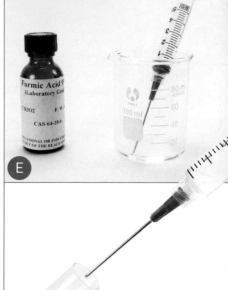

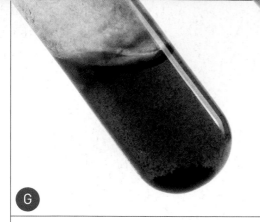

G

H

I

3. FILTER THE INK

After 12 hours, the solution should look clear with gray sediment of silver particles in the bottom (Figure **H**). In order to use the ink in an inkjet printer or airbrush, these particles must be filtered out to avoid clogging.

Remove the plunger from a new dispensing syringe, and place a 0.2μm syringe filter onto the syringe tip. Fill the syringe with the prefiltered solution, and replace the syringe plunger (Figure **I**).

Pressing slowly but firmly on the plunger, force the solution through the filter and into a small glass vial for storage (Figure **J**).

4. USING THE INK

Before using the ink, it's important to choose a suitable material to deposit it onto. In order to boost the conductivity of the ink, it must be heated to 90°C (192°F), so any material you choose must be able to withstand at least that much heat. Using this ink on porous materials such as paper or fabric will not result in a conductive coating, so it is recommended that

the material have a smooth surface.

Using a paintbrush, apply the ink to your material of choice; a stencil can be used to create complex patterns. Allow the ink to dry until it turns to a dull gray color.

Heat the material to 90°C (192°F) for at least 15 minutes. You can use a toaster oven or hot plate as a heat source.

> **CAUTION:** Heating in a toaster oven will make the oven unsafe for food preparation. ⚡

TIPS FOR USE

The ink is quite fragile and will easily scratch off most materials. To increase adherence, scuff the target surface before ink application. Once the ink is dry, apply clear nail polish to protect the traces.

Unfortunately, you can't solder onto the dry ink, as molten solder will leach the silver coating away. If you have a CircuitWriter silver-based ink pen, you can use it to create pads on the traces, which can then be carefully soldered to. ⊘

For complete step-by-step instructions and photos see makezine.com/diy-conductive-ink.

J

MAKE:
A FASHION STATEMENT

Wearable, interactive e-textile projects are a stitch away with supplies and kits from the Maker Shed.

ELECTRIC PAINT

BARE CONDUCTIVE

Bare Conductive Electric Paint

Draw—yes, *draw*—circuits with the ease of a glitter glue pen. This is the first electrically conductive nontoxic paint available today. Bare Conductive Electric Paint can be used on almost any surface: textiles, cardboard, paper, wood, and some plastics.
MAKERSHED.COM/ELECTRICPAINT

ProtoSnap LilyPad Development Board

This wearable e-textile development kit includes everything you need to get started, down to the needle and thread. The Lilypad was designed to be sewn into clothing and is even washable!
MAKERSHED.COM/LILYPAD

>MAKERSHED.COM

Maker Shed

The official store of **Make:**

UNDER THE HOODIE

An illustrated guide to wearables   WRITTEN BY GRETA LORGE

BODIES AREN'T STATIC, THEY DON'T HAVE STRAIGHT LINES, and after a while they tend to get dirty. So wearable systems embedded in garments and accessories have to be robust, flexible, and, ideally, washable (or at least removable). Here's a look under the hood — or hoodie, as it were — at the main components of wearable devices.

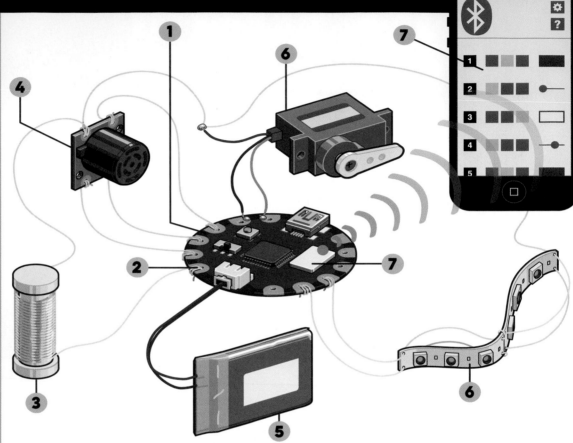

Rob Nance

1. CONTROL

Wearable-specific microcontrollers are small, so as to be comfortable and discrete. On the other hand, the distinctive shapes and colors can function as a decorative element. Several of the boards available are hand-washable (minus the power source). Read the documentation carefully.

2. INPUT/OUTPUT

In place of pins, these boards have metal eyelets which you can loop conductive thread through to sew soft circuit connections. Some boards also have snaps — or eyelets large enough to solder on snaps — for easy removal.

3. CONDUCTIVE TEXTILES

A material containing metals, such as silver or stainless steel, through which an electrical current can flow is said to be conductive. Wearable systems can make use of these materials in a variety of ways, such as:

- Thread for making circuits
- Fabric for capacitive touch sensors
- Hook-and-loop for switches

4. SENSORS

Sensors gather information about the environment, the user, or both. Examples of the former include light, temperature, motion (ACC), and location (GPS). Examples of the latter include heart rate (ECG), brain waves (EEG), and muscle tension (EMG). A few wearable microcontrollers have basic sensors onboard. Other manufacturers offer a range of external sensor modules that connect to the main board.

5. POWER

When scoping out a wearable design one of the first things to consider is the power requirement. Do you just want to illuminate a few LEDs, or do you want to run a servomotor? Boards with an integrated holder for a lithium coin battery are nice for low-power projects that need to be self-contained. However boards with a standard JST connector (with or without a circuit to charge LiPo batteries) are more versatile.

6. ACTUATORS

One generic way to describe a wearable system is: In response to X, where X is the input from a sensor, Y happens. Actuators such as LEDs, buzzers or speakers, and servomotors are what make things happen.

7. NETWORKING

To communicate with smart devices, the internet, or other wearable systems, you need wireless connectivity. In addition to wi-fi and Bluetooth, wearable-friendly options include:

- BLE, which has lower power consumption than classic Bluetooth, a range of 50m, and a data transmission rate up to 1 Mbps
- NFC, a radio frequency field with a range of approximately 20cm and data transmission rate up to about 400 Kbps ✪

BODY BOARDS
A guide to wearable microcontrollers

WRITTEN BY BORIS KOURTOUKOV

Created by Leah Buechley of MIT, and introduced commercially in 2007, the LilyPad was the first board to feature sew-through contacts for stitching soft circuits. Now there's a plethora of options in "ready-to-wear" microcontrollers. Here's a look at a few of the standouts.

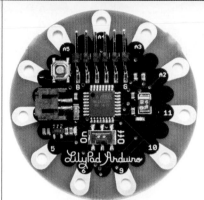

LILYPAD ARDUINO
- $20–$30 sparkfun.com/categories/135
- Dimensions 2" dia.

From the easy-to-sew tabs to the fact that it is washable, the **LilyPad** remains a great choice for projects that need to work with e-textiles. There are four core boards. With 22 pins, six analog, the Main board is handy if you need a lot of inputs and outputs for your project. (The other three boards have 11 pins, four analog.) The Simple and SimpleSnap are the same board as far as code is concerned, but the Snap has a permanently attached LiPo battery and snap tabs that allow it to be easily removed. The USB version has a LiPo charge circuit. The real power of LilyPad is an ecosystem that features numerous sensors and outputs, from switches to XBee breakouts.

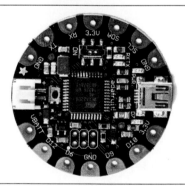

XADOW
- $20 makezine.com/go/xadow
- Dimensions 1" × 0.8"

Seeed's **Xadow** has a diverse selection of compatible modules, from a barometer to a UV sensor to a full GPS antenna. Thin, pliable connectors used to attach the various modules retain flexibility while maintaining stability.

FLORA
- $20 adafruit.com/product/659
- Dimensions 1.75" dia.

With design values similar to the LilyPad, the variety of sewable modules available for the **Flora** — from high-end GPS antennas to color light sensors — gives the system an edge when complex functionality is required.

Adafruit also makes the 1" diameter **Gemma**, at $8 a great option for small permanent projects that don't require a lot of I/O. An upcoming Arduino IDE-supported version will feature an on/off switch and micro USB connector.

TINYLILY MINI
- $10 tiny-circuits.com/products/tiny-lily.html
- Dimensions 0.55" dia.

The **TinyLily Mini** from TinyCircuits puts surprising power and I/O options into a dime-sized package. It's compatible with LilyPad and Arduino, however a micro-USB adapter is required to upload sketches.

SQUAREWEAR
- $22 rayshobby.net/cart/sqrwear-20
- Dimensions 1.7" × 1.7"

The open-source, Arduino-compatible **SquareWear 2.0** packs a lot into a compact footprint. Rather than having an ecosystem of swappable modules, it's intended as an all-in-one solution. It has a built-in mini USB, rechargeable Li-ion coin battery, and onboard light and temperature sensors.

More wearables kits and components available in Maker Shed makershed.com

ONES TO WATCH

PRINTOO
- $26 ynvisible.com/printoo
- Dimensions 1.38" × 1.38"

Printoo is currently the only platform for wearables that appears to have a robust flexible form factor. The e-ink-like 8-segment display is unique in the maker area, as is the flexible paper-thin batteries and conductive ink adapter they plan to offer. These will be really helpful in more discrete projects.

BITALINO
- $99 bitalino.com
- Dimensions 2.5" × 2"

BITalino is currently available as a $185 all-in-one board (4.5" × 2.5") with ECG, EMG, EDA, accelerometer, and light sensors or as a $197 "freestyle" kit with the individual biosensing modules detached. But the people behind it have launched another Kickstarter campaign for an iteration they're calling BITalino (r)evolution that they say will be smaller, more affordable, and more customizable.

FASHION

Nine designers blending apparel and technology

WRITTEN BY MATT RICHARDSON

IN FASHION, THE WORK THAT YOU SEE ON THE RUNWAY IS OFTEN A CONCEPTUAL GLIMPSE INTO THE FUTURE OF THE INDUSTRY. The same can be said for how makers use technology: The projects they create can anticipate where consumer technology is headed. It's even possible that maker projects can influence future consumer products.

At Wearables on the Runway at Engadget Expand in November, we saw that fashion and making go hand in hand. With the help of wearables experts Becky Stern and Kate Hartman, *Make:* editors showcased some of the best conceptual wearable technology from top designers. ◉

Elizabeth Tolson created these LED-adorned ballet dresses for Arch Contemporary Ballet.

Log(me) by Michelle Cortese listens to you and offers emotional scores.

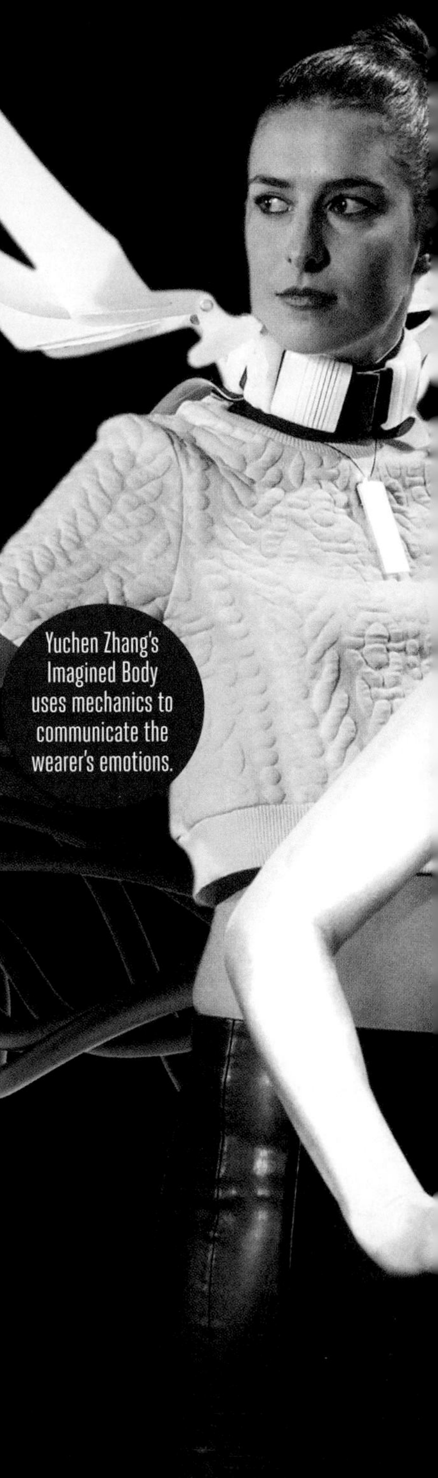

Yuchen Zhang's Imagined Body uses mechanics to communicate the wearer's emotions.

Want to make your own version of these wearables projects, or something totally new? Grab the following parts from **Maker Shed** at bit.ly/makewear to get started:

AUGMENTED SKIRT:
Flora GPS starter pack: #MKAD51
Servo: #MKPX17

MEU:
NeoMatrix: #MKAD73
Arduino Mini: #MKSP17

BALLET DRESSES:
LilyPad kit: #MKSF9

OTHER WEARABLE KITS:
Accelerometer: #MKPX7
Arduino Uno Rev 3: #MKSP99
XBee wireless starter kit: #MKPX19

FORWARD

Sebastian Guarin and Robert Tu collaborated to create a series of LED-enhanced garments.

Frog design's conceptual work envisioned life with wearable drones.

Som Kong integrated Robert Tu's MeU LED matrix for this overcoat.

This feathery skirt by Birce Ozkan points the wearer towards north.

x.pose by Xuedi Chen & Pedro Oliveira exposes skin based on online activity.

Wearable technology experts Kate Hartman and Becky Stern joined *Make:* contributing editor Matt Richardson to offer commentary on each of the garments. Matt is sporting Macetech's LED Matrix Shades (see page 59).

How to select and use sensors in wearable electronics

WHAT TO SENSE

Written by Kate Hartman ■ Illustrations made with fritzing.org

KATE HARTMAN is an associate professor at OCAD University in Toronto where she leads a research and development team dedicated to exploring body-centric technologies in the social context.

More wearables kits and components available in **Maker Shed** makershed.com

IT'S EASY TO HEAR ABOUT A COOL SENSOR AND DECIDE TO DO A PROJECT WITH IT.

"Oh, there's a really neat X sensor that just came out. I should obviously do an X project!"

But this leads to an interaction that's designed around the technology rather than *technology that's designed around a particular interaction.*

When working with sensors, a good place to start is to think about *what you're trying to sense.* What is the motion, action, or condition? What is the context and environment? What are the important aspects to consider? Then you can ask questions like these:

- "What different sensor (or sensors) could I use?"
- "What do I want to measure?" (sound, light, pressure, presence, etc.)
- "Where should the sensors live?"
- "What should I be looking for in the data I am gathering from them?"

In the following excerpt from the Sensors chapter of *Make: Wearable Electronics,* you'll look at some possibilities for what to sense and a starting selection of sensors that will fit the bill. But keep in mind that this is just the tip of the iceberg. Once you have a project idea in mind, you should go out and research what's available to best help your idea come to life.

FLEX

Bodies are bendy and it just so happens that *flex sensors* sense a flex or a bend. They're very good for areas of the body that bend in a broad, round arc. They work well on elbows, knees, fingers, and wrists. They are variable resistors and need to be used in combination with a voltage divider circuit in order to be read by a microcontroller.

FIGURE 1 : *Flex sensor circuit diagram*

The biggest challenges in working with flex sensors are positioning and protection. In order to get an accurate reading of the flex of your elbow, the sensor needs to be

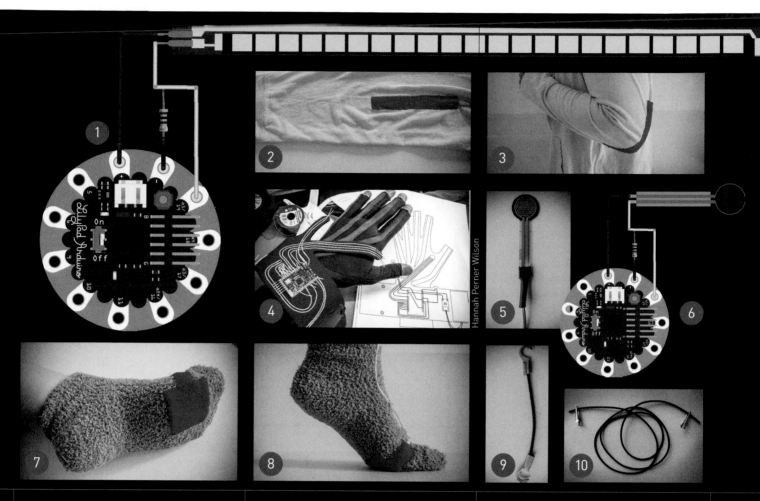

Hannah Perner Wilson

positioned in the same place on your elbow every time. Creating a secure pocket for the sensor can help with this.

FIGURE ②: *Sleeve to hold flex sensor in place*

FIGURE ③: *Flex sensor on a bent elbow*

The other thing to consider is that while flexing is a pretty rigorous and strenuous activity, many flex systems are fairly delicate, particularly at their connection terminals. Be sure to protect your connections. Reinforce with heat-shrink, and protect them with some sort of material.

FIGURE ④: *"The Gloves Project" uses flex sensors to create experimental gestural music; the project is developed by Rachel Freire, Imogen Heap, Seb Madgwick, Tom Mitchell, Hannah Perner Wilson, Kelly Snook, and Adam Stark.*

FORCE
Bodies often touch and get touched. One way to sense touch is through the use of *force-sensing resistors* (*FSRs*). FSRs have a makeup that's similar to flex sensors but are configured to be sensitive to pressure rather than bending. They are also variable resistors and have delicate connections similar to flex sensors.

FIGURE ⑤: *FSR with heat-shrink tubing used to protect the delicate solder connections between the sensor and wires*

FIGURE ⑥: *FSR circuit diagram*

FIGURE ⑦: *A pocket sewn onto the sock helps keep the FSR securely in place.*

FIGURE ⑧: *When pressure is applied to the ball of the foot, the change in sensor data can be read by the microcontroller.*

STRETCH
From the bend of a knee to the expansion and contraction of a rib cage with each breath, properly positioned stretch sensors can capture the fluctuating nuances and

curves of the human form. A stretch sensor is simply a conductive rubber cord whose resistance decreases the more it gets stretched. This is yet another example of a variable resistor.

FIGURE ⑨: *Stretch sensors with hooks attached*

Stretch sensors come precut at different lengths with hooks crimped to either end for easy connection, or you can buy it by the meter and cut it to whatever length you need .

FIGURE ⑩: *Stretch sensor with hardware for customization*

Stretch sensors are a fun material to work with. They can also be elegantly incorporated into textiles through knitting or weaving.

FIGURE ⑪: *"Aeolia" by Sarah Kettley, with Tina Downes, Martha Glazzard, Nigel Marshall, and Karen Harrigan, explores the process of incorporating stretch sensors*

11

Tina Downes and Catherine Northalt

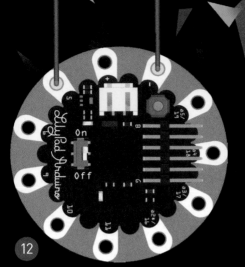

12

13

14

15

Collin Cunningham

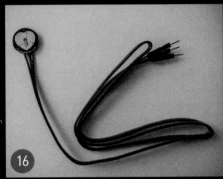

16

into garments through weaving, knitting, and embroidery techniques.

MOVEMENT, ORIENTATION, AND LOCATION

People are active and mobile creatures. They reach for things they want, turn toward loud noises, and crouch down to coax the cat from under the bed. When creating wearables that react to events such as these, it is helpful to be able to sense movement.

A cheap and easy way to sense movement is through the use of tilt switches .

FIGURE 12: *A basic tilt switch can be read by a digital input pin.*

But there are also far more sophisticated sensors that you can use. *Accelerometers* measure acceleration or changes in speed of movement. They can also provide a good measurement of tilt due to the changing relationship to gravity.

Accelerometers have a set number of axes — directions in which they can measure. The ones shown above are three-axis accelerometers, meaning they can measure acceleration on the X, Y, and Z plane.

FIGURE 13: *Accelerometers: LilyPad Accelerometer, ADXL 335, Flora Accelerometer*

FIGURE 14: *The accelerometer shirt by Leah Buechley uses accelerometer data to control the color of an RGB LED.*

If tilt, motion, and orientation aren't enough, and you want your wearable to know where you are on the planet, GPS is the way to go. Just like your car, bike, or phone, your jacket or disco pants can have GPS, too. There are a number of Arduino-compatible GPS units available, but the Flora GPS is a compact and sewable option.

FIGURE 15: *Flora GPS Jacket by Adafruit, Becky Stern, and Tyler Cooper*

HEART RATE AND BEYOND

Your heart beats faster when you're excited, and your skin gets clammy when you're nervous. Besides sensing your environment and your movements, you can also use sensors to learn more about what is happening within someone's body. A great place to start sensing these biometrics is pulse or heart rate.

Optical heart rate-sensors, such as the Pulse Sensor Amped, are a small, lower-cost solution for measuring pulse. This type of sensor measures the mechanical flow of blood, usually in a finger or earlobe. It contains an LED that shines light into the capillary tissue and a light sensor that reads what is reflected back. It produces varying analog voltage that can be read by the analog input on any Arduino.

FIGURE 16: *Pulse sensor*

FIGURE 17: *Pulse sensor circuit diagram*

Chest-strap heart monitors are a more expensive but more accurate solution for measuring heart rate. They measure the actual electrical frequency of the

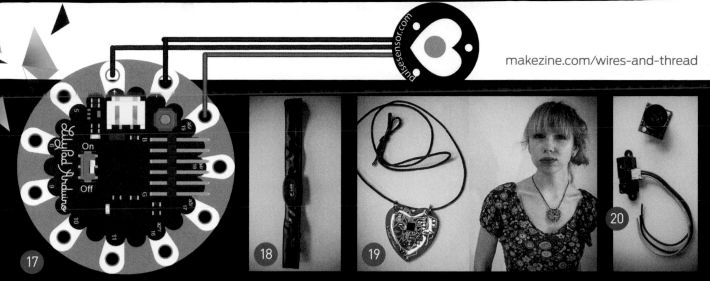

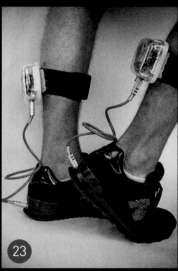

heart through two conductive electrodes (oftentimes made of conductive fabric) that must be pressed firmly against the skin. Polar produces heart rate monitors that wirelessly transmit a signal with every heartbeat.

FIGURE 18 : *Polar heart rate monitor band*

FIGURE 19 : *"Heart Spark" by Eric Boyd is a custom-designed, printed circuit board necklace that receives a signal from a Polar heart rate monitor band and blinks in unison with the wearer's heartbeat.*

PROXIMITY
Sometimes you will want to know how close or far away something is from the body. Proximity sensors are useful for detecting nearby objects, walls, or even other people. When selecting a proximity sensor, it is worth considering what your desired range or detecting distance is, as well as what sort of beam width you need to monitor.

FIGURE 20 : *Ultrasonic and infrared proximity sensors*

FIGURE 21 : *"Augmented Vision" by Greg McRoberts is a wearable seeing aid device that uses flashing RGB LED to represent fluctuating data gathered by an infrared heat sensor and ultrasonic distance sensor.*

LIGHT
The most basic type of light sensor is the photocell. Its resistance varies based on the level of light it senses. Some have resistance that increases as the light level increases, but some have the reverse relationship. This can be quickly determined by viewing the sensor values in the serial monitor.

FIGURE 22 : *Photocell, LilyPad Light Sensor, Flora Light Sensor*

The photocell is a great sensor to work with because it is small, easy to manipulate, and incredibly inexpensive. It can be used to sense ambient light levels, but it can

also be used for less intuitive purposes like determining whether a jacket is open or closed or if the heel of a shoe is on the ground or in the air.

FIGURE 23 : *"Perform-o-shoes" by Andrew Schneider are music-controlling footwear that have a photocell embedded in the bottom of the heel; the higher the shoe is off the ground, the faster the music will play.* ⊘

Make:

Wearable Electronics
Design, prototype, and wear your own interactive garments

✚ This is a condensed excerpt from the Sensors chapter of *Make: Wearable Electronics: Design, prototype, and wear your own interactive garments.* Available in the **Maker Shed** makershed.com/products/make-wearable-electronics

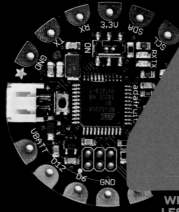

4 FUN FLORA PROJECTS

Bring your sci-fi fantasies to life with this sewable board

WRITTEN BY
LESLIE BIRCH

THINK BACK TO YOUR FIRST WEAR-ABLE TECH EXPERI-ENCE: maybe it was the communicator badges in *Star Trek*, the palm flower crystals in *Logan's Run* or the man-turned-machine in *Terminator*. These devices could simplify communication, monitor life functions, or even transform physical bodies. We all loved it, and we're still seeking it. We want science fiction to be real because we want to feel special, aware, and powerful. That's why wearable tech is our now, and the Flora microcontroller is our BFF.

Flora is a tiny but mighty microcontroller from Adafruit that is stitch- and solder-friendly (see page 35). It's got a built-in USB jack for easy Arduino programming and also a JST jack for fast battery plug-in. The best part is this microcontroller loves NeoPixels. If you haven't seen NeoPixels, imagine mega-bright RGB LEDs with the ability to talk to each other. Adafruit's NeoPixel library for Arduino has made this an easy task, so anyone can make a wearable light show.

So, what do you want to be? What will be your transformation? Check out these four fabulous projects and make your own sci-fi real life. ◗

LESLIE BIRCH is a tech geisha, with a love of open source hardware — especially Arduino. She's crafted award-winning wearables and currently creates projects for Adafruit and Element 14. Follow her @zengirl2

SOUND REACTIVE HEADPHONES

There's a music duo that just beats them all when it comes to nerdy chic tech. They're French and they got popular in the late '90s — you had better be thinking Daft Punk or no Quizzo for you. These headphones by caitlinsdad are like a light organ for the ears, making you an instant Daft Punkee.

This mod is a combo hack of two Adafruit tutorials, and it's sweet! The headphones use a Flora and two NeoPixel LED Rings facing the speakers for the light show. Not only does this soften the lighting, but it creates a mysterious effect. Meanwhile, the microphone amplifier board works its magic syncing the LEDs to the music. You should use a soldering fan for this project, and be prepared to tidy up the wires at the end. It's a great build for your listening pleasure — or not, as a friend will probably steal them from you. (makezine.com/go/light-up-headphones)

SUPERHERO ACTION BANDS

What's one thing all superheroes have in common? They've got a special power, or at least a device that makes them boss. These action bands by designers Niki Selken and Annelie Koller will make you the tower of power. Move your arm and show off your color-changing lights and sounds. Feel free to sock like Batman, chop like a Ninja Turtle, or ward off bullets like Wonder Woman. Your powers are about to be known.

These bands are a family-friendly project involving some sewing and light soldering. The fabric base is a hacked headband with decorative laser-cut pieces (which can be simplified for young ones with felt scraps). A Flora reads an accelerometer to capture the arm movement, which is then sent to a magical NeoPixel and piezo speaker for heroic fun. Have your cape and video camera ready. (wearabytes.com/superhero-action-bands)

Annelie Koller and Niki Selken

Hep Svadja

DISCO DRESS

You've always wanted Lady Gaga's futuristic fashions, and now you don't have to wait. This dress by maker Samuel Clay has 40 dancing LEDs that are sure to get attention. Will your light pattern be hula-hoop, sparkles, raindrop, spiral, or random colors? Let your emotions be your guide — you're sure to look glam with any of them.

This dress-hacking project is for someone comfortable with sewing, as you'll need to stitch six LED strips to the lining of a dress. If you're crafty, you can also choose to create casings of sheer ribbon for the strips. For the wiring inside, you may want to consider Adafruit's incredibly flexible silicone wire. Once you've got everything assembled, just program the Flora with your favorite pattern and you'll be off dancing in no time. You go, modern Marilyn Monroe. (makezine.com/go/light-up-dress)

Sandra Krampelhuber

ELECTROMAGNETIC FIELD DETECTING DRESS

We live in a complicated world surrounded by devices and EMFs. Maybe you're starting to question your environment and fancy yourself needing a spacesuit for planet Earth. If so, this is the build for you. Designed by Afroditi Psarra, this dress allows you to sense EMFs through haptic and sonic feedback. The most interesting thing about the dress is its sensors: two embroidered copper coils that act like antennas at the wrists. You are the detector.

The Flora takes in information from the wrist antennas and translates that as an output using a vibration motor and headset. The zipper is more than just decoration, providing a filter and volume control for the audio. This project requires a bit more advanced sewing and soldering expertise, but when you're through, you'll know more about your neighborhood than the Army Corp of Engineers. Be all you can be. (afroditipsarra.com)

THE FASTEST MAN ON EARTH

How I built Bionic Boots to run like an ostrich

WRITTEN BY KEAHI SEYMOUR

KEAHI SEYMOUR
Originally from Solihill, England, Keahi Seymour tends bar in San Francisco, but his passion is design and inventing. Someday, he wants to leverage these skills to build a low-cost, portable, water distillation and purification device for developing nations.

TO STRAP ON AND RUN WITH THE BIONIC BOOTS IS A FEELING LIKE NO OTHER.
As you begin to stride, you feel the springs storing energy. Then you push off, and you feel the enormous power released, akin to acquiring your own slice of a superpower.

I built the Bionic Boots simply because I wanted that experience. Ever since I was 12 years old, I've been dreaming of one day dropping into the African savanna and running with cheetahs.

Initially, the concept was to emulate and experience the sensation and speed of running like a fast animal. I still have the same goal, but the invention is evolving through the use of future technologies to reach new endeavors. I want to produce a viable form of environmentally sound transportation over any terrain, be it city streets or off-road trails — and to run faster than any man alive.

The first spark of inspiration came from watching a natural history program on kangaroos and how they were able to store energy in their large Achilles tendons, enabling them to move at high speed over difficult terrain with an efficient gait. I made my first drawings at 12, and the inventions that came along years later were not too dissimilar to those originals.

A quarter century later, I'm on something like the 200th prototype. These boots are made from aluminum and carbon fiber, with elastic tendons. In them, I stand 7 feet tall, and can run 25 mph.

In the intervening time, I went to college and moved, in 1999, to America. I studied transport design and won a Royal Society for the Arts award for the boots, and used that grant to go to California, the birthplace of so many recreational sports (mountain bikes, skateboards, fiberglass surfboards) and to bring the Bionic Boot into the public domain.

I already had an aesthetically pleasing and working prototype, but it would be one of many. I tend bar six days a week to pay for patent fees and material costs and have worked up to the X14 (2014) prototype.

MECHANICAL ANIMAL
The boots work by basically giving plantigrade (that is, flat-footed) humans a mechanical advantage, allowing them to run on their toes in digitigrade fashion, the way fast land mammals such as greyhounds and cheetahs do. By raising me on my toes, the boots lengthen my legs and stride, which increases speed and efficiency.

There are two levers, a main one and one for the toes. Both are attached to rubber extension springs that mimic the aforementioned kangaroo tendons. The

Hep Svadja

main lever provides the majority of the propulsive force. As the boot lands, the 18-inch lever stretches the springs; then as they contract, the lever swings through a pivot past the heel, propelling the main lever and springing the user like a catapult. The smaller toe lever has a rubber and foam grip, which gives purchase over uneven terrain. Depending on the conditions, one, two, or possibly three interchangeable toes with differing sizes and tread patterns can be added.

To build the boots, I leveraged metalwork, carbon-fiber molding, and spring building. The initial main boot was designed and constructed by making an anatomically correct copy of the boot itself. Later, my friend Carl Riccitelli made a mold of it, and laid carbon fiber into the mold to produce the current prototype with the best strength-to-weight ratio so far.

Aircraft grade aluminum (6061 and 7000 series) was used for the other major components. All were constructed without the use of CNC milling machines or casting methods, but instead were cut, shaped, and polished using only a hand drill, angle grinder, and hacksaw.

The spring system is made of natural rubber from speargun spring bands, cut to specific length, with custom-made grommets to attach to the rubber. These can be added or subtracted to adjust for the weight of different users or the running style or cadence desired.

My invention has been designed — and has evolved — for fast running as a form of transportation. While the top speed is formidable, the stilts are not designed for maneuverable running (i.e. turning).

I am not the only inventor to design augmenting boots. There are other prototypes and products that use different spring systems, though my invention pre-dates the patents on the most similar ones.

One example, originally sold as Powerbocks (now Pro-Jumps) from Germany, also uses a pivoted lever to add stability, but is designed more like traditional stilts, allowing you to jump vertically like a pogo stick. Young athletes use them for parkour-style extreme sports. (Not to be confused with kids' springy Moon Shoes, which simply suspend the feet in oval frames, like two trampolines.)

The differences between the two are not only in the main function but also in the spring: The Pro-Jumps use a fiberglass leaf spring, whereas mine feature natural rubber as an extension spring to store potential energy. Additionally, because of the material, the Bionic Boots weigh in at just six pounds, around two pounds less than the Pro-Jumps.

But one feature that is truly unique is the pivoted toe, which gives the landing some dampening from the stiff impact of the large spring, as well as maneuverability and extra purchase on uneven terrain. I've used them on everything from potholed and cobblestoned streets in New York and London to California beaches, peaks in the Rocky Mountains, and even in shallow water.

> I've used them on everything from potholed and cobblestoned streets in New York and London to California beaches, peaks in the Rocky Mountains, and even in shallow water.

BIONIC BOOT EVOLUTION

X5-2005 — All-aluminum prototype using a snowboard binding to hold the leg, a tubular lever, and a single toe. Spring is rubber tubing, a progression from the original bungee-cord spring.

X8-2008 — First use of carbon fiber on the exoskeleton. At 17 inches, this prototype is larger with a more powerful spring and longer stride, but upgrades increased the weight.

X10-2010 — All-aluminum prototype with decreased height for increased maneuverability. Decreased ride height also decreases the amount of power added to the spring.

FARTHER AND FASTER

There's still a lot to do, future improvements to the Bionic Boot to extend the distance and speed. I'm planning an onboard electronic feedback control system to help coordinate the power and propulsion to give the most effective timing of power output throughout the running cadence, thus providing maximum efficiency and power expenditure. I'd also like to explore 3D printing, specifically with titanium or carbon fiber — even a 10 percent weight reduction could give incredible results. Collaboration with companies like Local Motors, which printed a car with carbon fiber-infused ABS, or Renishaw, which printed a titanium bicycle, could help improve the boots' performance.

The "muscles" could be adapted too, perhaps to include pneumatics like Festo's "fluidic muscle," which enabled that company's Bionic Kangaroo. It uses pneumatic pressure to contract the muscle as air is added. In nature, a kangaroo recovers energy from jumping and stores it for the next leap. In a boot, that could mean greatly increased speed and distance.

In the end, the Bionic Boots could become a whole interlinked exoskeleton built solely for speed, approaching that of an ostrich or even a cheetah.

I can see a vision of a prototype I sketched many years ago, encompassing a full-powered protective suit with onboard readouts of speed, distance, system power outputs, and more. It's my bionic conception of future transportation. ❷

The Bionic Boots could become a whole interlinked exoskeleton built solely for speed, approaching that of an ostrich or even a cheetah.

X10 (rev.)-2010 — Altered the exoskeleton shape. More advanced structure doesn't use aluminum tubes for main support — the sculpted metal provides the strength.

X11-2011 — First fully carbon-fiber exoskeleton, molded to encompass the anatomy of my leg. The shape transmits the force in the boot as the main lever stretches the spring.

X12- 2012 — A carbon-fiber lever and toe reduced weight but failed due to the high forces involved when running. Double toe increased mobility on uneven terrain.

X14- 2014 — Longer rubber springs to increase power. Bored-out aluminum lever provides a strong, reliable mechanical component. Reinstated lighter single toe.

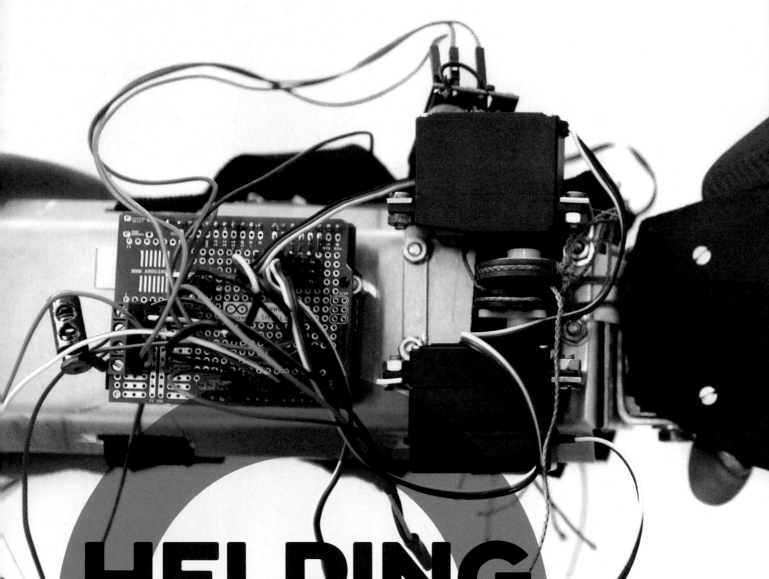

HELPING HANDS

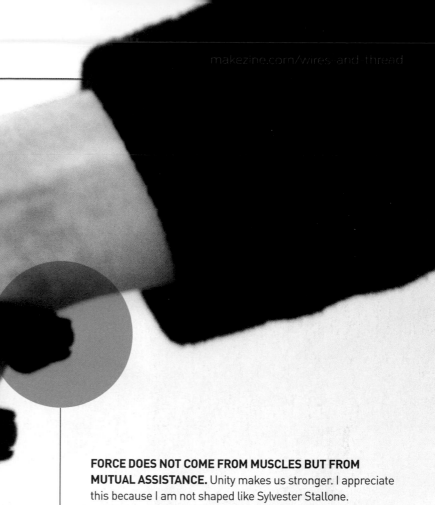

FORCE DOES NOT COME FROM MUSCLES BUT FROM MUTUAL ASSISTANCE. Unity makes us stronger. I appreciate this because I am not shaped like Sylvester Stallone.

Since an accident with a hydraulic press 12 years ago, I wear a myoprosthesis — an electrical hand that I can control by muscle sensors. There are new models that are much better, but they are very expensive. I tried them all –from RSLSteeper's bebionic to Ottobock's Michelangelo — but even $50,000 models were lacking. It was a moot point; my favorite was not covered by health care.

At first, I thought this was unfair, that I would not benefit from the articulated, carbon-fiber fingers, the variable speed, 14 grip patterns, and myoelectric sensors on the bebionic. But then I remembered how lucky I was to have the version I had, thanks to decades of research and access to first-world health care. What about people without that access?

HOW I BUILT
A 3D-PRINTED
PROSTHETIC HAND
— FOR MYSELF
BY NICOLAS HUCHET

Miguel Templon

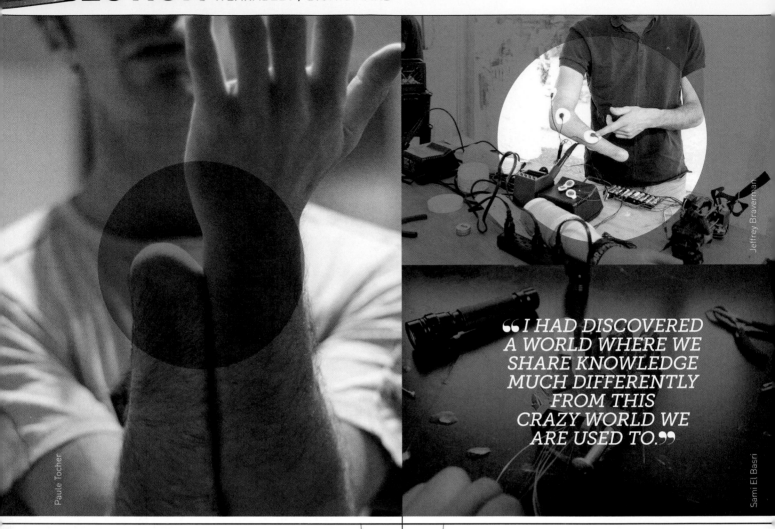

I HAD DISCOVERED A WORLD WHERE WE SHARE KNOWLEDGE MUCH DIFFERENTLY FROM THIS CRAZY WORLD WE ARE USED TO.

Paule Tocher

Jeffrey Braverman

Sami El Basri

There are a few options. Alternative prosthetics — many 3D-printed, articulated, and open-source — built at universities, fab labs, and even companies, are beginning to fill the gaps. California-based Not Impossible Labs built a lightweight 3D-printed hand based on the open-source Robohand, which itself has been downloaded more than 9,000 times from Thingiverse. A loose coalition of volunteers called e-NABLE includes Robohand creators Ivan Owen and Richard van As, and boasts eight different designs as well as video tutorials for making them. Others are based on the hands of InMoov, the open-source, 3D-printable robot from France. It's the basis for a design by Ananya Cleetus, a 17-year-old girl from Pennsylvania, for Help is at Hand, a robotic prototype she brought to the White House Science Fair. The Open BioMedical Initiative is working on several open-source hands of its own, and the Open Prosthetics Project shares dozens of designs, how-tos, and advice.

The list goes on. But I didn't know about any of them. I didn't even know about 3D printing.

In October 2012, while walking through Rennes, France, where I live, I passed an exhibition where strange machines, like something from science fiction, were depositing layers of material onto platforms. They were 3D printers.

I entered, asking "Excuse me, is it possible to make a robotic hand with this? Because I have a prosthetic hand." Usually when people see my disability they try not to look at it or ask what happened, but their reaction was different. They were excited. They wanted to know how it worked. "We can download a robotic hand on Thingiverse and make the pieces with this 3D printer," one said. "Here is one called InMoov, the guy is using Arduino and it's open source."

I didn't understand a word they were saying. But I knew what they meant: It's possible to design an inexpensive bionic hand that you can make yourself, then share your work so other people can improve it and share it further. I had discovered a world where we share knowledge much differently from this crazy world we are used to. I was looking at things differently; it was my revolution, my change.

We launched the project a few months later, in February 2013, with the LabFab in Rennes. But it was a truly international effort: printed digits from Rennes; muscle sensors from the U.S.; and design input from Brazil.

The foundation is a 3D-printed hand, equipped with actuators to move the digits and joints, fishing line to connect the actuators to those joints, muscle sensors and a socket where

Thomas Mortier

Miguel Templon

Sami El Basri

they rest, batteries, and an Arduino brain. It's all built for a price around $250.

That's a vastly lower sum than typical hands, which run from $8,000 to $80,000. For decades, a German company called Ottobock dominated the relatively small market. Touch Bionics appeared in 2003, and it built i-limb, the first prosthetic hand with articulated digits. Then RSLSteeper released the first of three versions of bebionic. DARPA pushed for advancements as well, due in part to a growing number of injuries from Iraq and Afghanistan. These devices have a few things in common, not least their articulation and multiple degrees of freedom.

For a low-cost hand to challenge medical-grade prosthetics, there are a few non-negotiable capabilities it must have. It has to be durable and reasonably lightweight, with an opposable thumb. It should reach a maximum open size at least large enough to grab a coffee cup or a water bottle, yet close precisely enough to hold a coin, a pen, or the cable to your headphones. It must grasp laterally — in a neutral position, like holding a key — and palm-down, as if to carry a suitcase. And it helps to design something fun to build and with an attractive aesthetic.

But first, we needed a prototype. Along with my collaborators

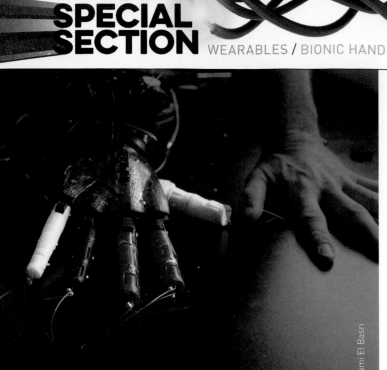

Sami El Basri

NICOLAS HUCHET
teaches audio,
including live sound
and Pro Tools. He also
makes music, plays
drums, and travels. If
you want to be part
of his international
network, visit
bionico.org

from LabFab Rennes, I presented the first prototype, with 3D-printed plastic, some fishing line, and an old Nesquik box, at the Tu Imagines? Construits! festival in Rennes in June, 2013, with Gael Langevin, the creator of InMoov.

From there, things took off. We got invited to the first European Maker Faire in Rome. I learned how to use a 3D printer and we made a second slightly improved prototype. We got invited to Geek Picnic, a science and technology festival in St. Petersburg, Russia, and brought a third prototype to World Maker Faire New York.

Again and again, we were asked the same question: What's that?

It's a device for which there is great need. In the United States, insurance may pay for a prosthesis. In Colombia, they give you a hook; and in Russia, not much else. Spain is really suffering — kids have access to prosthetics, but often, adults won't because of the price. And that's not even mentioning emerging countries.

Bionico is not yet robust enough to be a prosthesis. It's a prototype, but with it we've traveled the world as self-financed volunteers. So what do we do now? We may organize a crowdfunding campaign. We may look for sponsors who share our philosophy of a utopia of health for everyone. Above all, we want to create an international network and database devoted to improving low-cost prosthetics. This is an open-source project, which means you can participate or make it yourself. The prosthetic-hand field is very small, but if we build a bridge between countries and people, we can make it better and stronger, and go further, faster. As the American philosopher Sylvester Stallone said, "Big arms can move rocks, but big words can move mountains." ◢

Miguel Templon

UNDERWARE
WRITTEN BY BO MOORE

Subdermal devices are the cutting edge in wearables

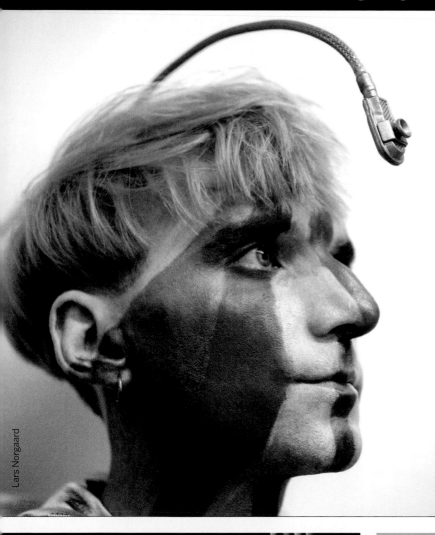

Lars Norgaard

SURE, WEARABLES ARE GREAT, BUT THEY'RE STILL EXTERNAL DEVICES — accessories we have to don, that we can lose, break, or have stolen. They do cool things, but don't inherently change anything about who we are. But there's a small community thinking beyond wearables. These self-described biohackers create DIY implantable systems to augment human capabilities and transcend the limits of biology.

Common entry-level biohacks include neodymium magnets embedded in fingertips to sense magnetic fields and pick up small objects and RFID or NFC chips in hands for unlocking phone screens or interacting with other compliant devices.

And then there are some individuals who are pushing even those limits.

HACKED SYNESTHESIA

Neil Harbisson was born colorblind. And not just mix-up-red-and-green colorblind; to him the whole world looks like a black-and-white TV — everything grayscale. But thanks to a snorkel-like device mounted to his head, Harbisson listens to color. A small camera translates the visible spectrum into sound — different frequencies for different hues — and a chip transmits it to him via bone conduction. The device is so integrated with Harbisson's person that he now hears color passively, even experiencing the sounds of colors in his dreams.

UNDER ARM

For 90 days in late 2013, **Tim Cannon** traveled internationally, exercised, was submerged in water, and led an otherwise normal life — all with a smartphone-sized device implanted in his forearm. It was the first version of Circadia. The device charged wirelessly, mimicked bioluminescence by lighting up subdermally, and sent Cannon's body temperature to his phone.

Now that he's proven it can be done, Cannon and the Grindhouse Wetwares biohacker community he works with are testing batteries and power sources to improve efficiency, and hoping to transmit other body metrics such as heart rate, blood oxygen level, and more.

Andrew Obenreder

IN-EAR HEADPHONES

Rich Lee has a magnet in his finger, but he wanted more. In 2013, Lee had two magnets implanted in his head, one in the tragus of each ear. Using a homemade amplifier worn around his neck, he can listen to music through the magnets. The signal from his MP3 player runs through a coil, creating an electromagnetic field that causes the magnets to vibrate, playing music.

Lee's coil works with almost anything you can plug a headphone jack into, like a metal detector. A piezo contact mic lets him "hear" through walls, and hooking his coil up to a rangefinder grants a crude version of sonar. Lee is also combining lie-detection, stress-analysis, and voice-recognition apps into a perception-enhancing tool. Take that, Spidey-sense!

Rich Lee

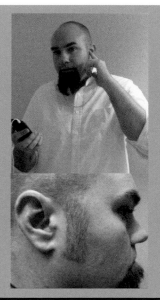

Build a low power watch from off-the-shelf components and breakout boards

OPEN-SOURCE

WRITTEN BY JONATHAN COOK

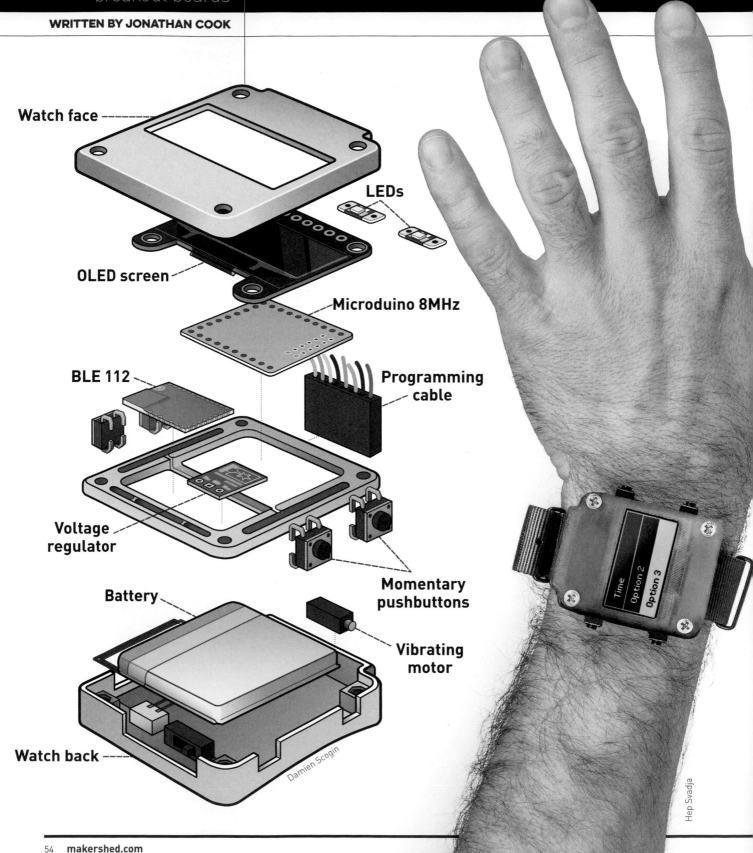

Watch face

LEDs

OLED screen

Microduino 8MHz

BLE 112

Programming cable

Voltage regulator

Momentary pushbuttons

Battery

Vibrating motor

Watch back

Damien Scogin

Hep Svadja

SMARTWATCH

TIME REQUIRED:
20–40 HOURS
COST:
$75–$125

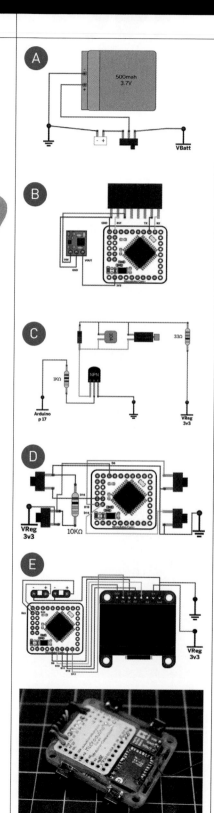

MATERIALS

- » **Microduino-Core+ microcontroller module,** 8MHz, 3.3V, ATmega644PA or ATmega1284P from microduino.cc
- » **Voltage regulator, 3.3V** Pololu #2114, pololu.com
- » **Motor, vibration** Pololu #2265
- » **Tactile switches, momentary pushbutton (4)** Adafruit #367, adafruit.com
- » **Screen, OLED** Adafruit #938
- » **Battery, LiPo, 3.7V 500mAh** Adafruit #1578
- » **Switch, power** Adafruit #805
- » **Connector, JST-PH 2-pin** Adafruit #1769
- » **Wire, fine** Adafruit #1446
- » **LED sequins** Adafruit #1758
- » **Charger, LiPo, Micro-USB** Adafruit #1304
- » **Jumper wires, female** Adafruit #266
- » **Jumper wires, male** Adafruit #758
- » **Bluetooth LE chip** Mouser Electronics #603-BLE112-A, mouser.com
- » **Watch strap** Amazon #WS-NATO-BB-22M, amazon.com
- » **3D-printed parts** Grab the files for free at oswatch.org/3d_printing_build.php.
- » **Resistors: 10kΩ, 1kΩ (1), and 33Ω (1)**
- » **Transistor, NPN**
- » **Capacitor, 0.1µF**
- » **Diode**

TOOLS

- » **FTDI Friend** Maker Shed #MKAD22, makershed.com, or similar FTDI-USB programmer
- » **CC Debugger** Texas Instruments #CC-DEBUGGER, to program the Bluetooth chip
- » **Computer, PC** Bluetooth programs are PC-only.
- » **Arduino microcontroller board** an extra, to burn the bootloader and debug
- » **Soldering iron, temperature-controlled**
- » **Solder with flux core**
- » **Tube of flux**
- » **Wire cutters/strippers, fine**
- » **Rotary tool**
- » **Sandpaper**
- » **Cyanoacrylate (CA) glue** aka super glue
- » **Epoxy, quick set**
- » **Multimeter**
- » **Tweezers**
- » **Solder sucker**
- » **Helping hands** aka third hand tool
- » **Breadboard**

MY OPEN-SOURCE SMARTWATCH combines readily available breakout boards, careful soldering, and a 3D-printed frame to make a one-of-a-kind timepiece that displays notifications from your smartphone and is easily customizable in function and appearance.

The watch design is straightforward, consisting of four major sections: a battery charging circuit, vibrating motor for silent alerts, programmable Arduino-compatible core with power regulation and Bluetooth LE, and an OLED display with pushbuttons.

Breadboarding the project is a snap. Wiring it into a small enclosure meant for the wrist is quite another matter. Break out your fine-point soldering iron and follow the complete instructions at oswatch.org.

BATTERY CHARGING
A 3.7V 500mAh LiPo battery is wired to a JST connector and a two-position switch. Switched to the right, the circuit is in battery mode. Switched left, it's ready for LiPo charging via the JST connector (Figure Ⓐ).

PROGRAMMABLE CORE
Within the 3D-printed frame an 8MHz Microduino microcontroller is connected to a programming port, a Bluetooth Low Energy board for communicating with your smartphone or other devices, and a voltage regulating circuit (Figure Ⓑ).

VIBRATING MOTOR
The simple vibrator circuit consists of a diode, 1K and 33Ω resistors, capacitor, NPN transistor, and motor. The circuit is then connected to the Microduino to buzz your wrist when new calls or alerts come in (Figure Ⓒ).

PUSHBUTTONS AND OLED DISPLAY
Four momentary pushbutton switches are wired to three pull-up resistors internal to the Microduino and a single external 10K pull-down resistor (Figure Ⓓ).

An OLED screen and two small LEDs are wired directly to seven of the digital pins on the Microduino to display time, text, alerts, and more (Figure Ⓔ). ◢

JONATHAN COOK
is a product manager by day, a hacker by night, and an artist when he has the time. He has been merging his love of technology & creativity since he could pick up a pencil and a soldering iron.

For complete step-by-step build instructions see makezine.com/open-source-smartwatch

CAPTIVATING COUTURE

WRITTEN BY CRAIG COUDEN

From mood-sensing displays to stunning light shows, the future of fashion is here

Illuminated clothing is one of the more, ahem, visible branches of wearable technology, so it's no surprise that a few feet of EL wire can turn heads and add an extra dimension to your outfit. But that looks downright old-fashioned compared to these fantastic light-up creations. Some of these projects show that less is more. And the rest show that *more* is more. Take note: It's not how bright you are, it's how stylish you look in the light. ◔

PROXIMA
Laura Dempsey, Hannah Newton, SAIT RADLab
makezine.com/go/proxima
Part wearable, part interactive duet, the female dancer's jacket is programmed to light up based on the proximity of RFID tags worn by the male dancer. The lights follow him as he moves around her — the closer he gets the more the LEDs sparkle.

TECH TIE 1.0
Jeff de Boer with Grant McKee and Shannon Hoover
makefashion.org/tech-tie
Tech Tie will definitely turn heads on the street. Sixteen small, OLED screens cycle through different animations controlled by a Seeed Xadow Main Board. Version 1.5 is on the way with e-paper screens, more functions, and smartphone control.

NEBULA PENDANT
Vlad Lavrovsky
makefashion.org/nebula
Why let the ambient light around you dictate how your jewels shine? Integrating conductive thread, high-efficiency LEDs, and BGA-interface integrated circuits, the pendant scatters colored light across the wearer's skin and clothing for a dazzling, unique piece of jewelry.

SYNAPSE DRESS

Anouk Wipprecht
vimeo.com/106431614

This 3D-printed dress measures attention levels and heartbeat through sensors attached to the wearer and reacts by lighting up LEDs. The data is logged along with video from an onboard webcam in order to review emotional triggers or moments of high focus at a later time.

GER MOOD SWEATER

Sensoree
sensoree.com/artifacts/ger-mood-sweater

Using galvanic sensors attached to the wearer's hands, the Mood Sweater reads excitement and translates into an "external blush" using color-changing LEDs in the collar.

Roger Dyckmans

Audrey Love

Yanaura

FIBER OPTIC DRESS

Natalie Walsh
Make it: instructables.com/id/Fiber-Optic-Dress

Inspired by jellyfish, this beautiful, bouncy dress uses 360 fiber-optic cables cut to different lengths to create the layered look. A kit is available online, or you can follow the online instructions and make it from scratch.

KINISI

Katia Vega
makezine.com/go/kinisi

Could your expressions act as an interface? A wink, a smile, or a raised eyebrow triggers sensors attached to various muscles and relays the information to a microcontroller that sets off different light patterns in LEDs attached to the face and hair.

Liteweave.com

ILLUMINATING FASCINATORS
LCH Designs
liteweave.com

All eyes will be on you when you hit the town — or the races — in a handmade, fiber-optic-equipped fascinator headpiece.

LumiLabs

DRAPER 2.0
LumiLabs
Make it: makezine.com/go/draper

This dapper pocket square from lighting design firm LumiLabs is an understated yet eye-catching homage to a classic style.

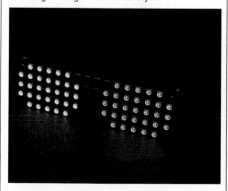

Garrett Mace

LED MATRIX SHADES
Macetech
macetech.com

Make Yeesus proud with this stylish piece of eyewear. Arduino-compatible with online documentation, program your own messages and patterns to suit your style.

GALAXY DRESS
CuteCircuit
cutecircuit.com/collections/the-galaxy-dress

Commissioned for the Museum of Science and Industry in Chicago, a single picture doesn't do it justice. Preprogrammed light patterns cascade through the 24,000 hand-embroidered, full-color LEDs to create the largest wearable display in the world.

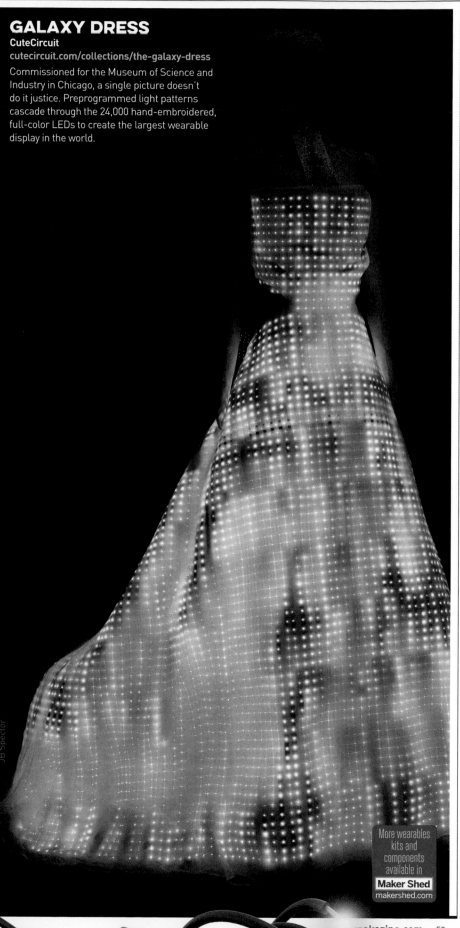

JB Spector

More wearables kits and components available in **Maker Shed** makershed.com

MIND-READING BEANIE

WRITTEN BY IO FLAMENT

Make an EEG hat that translates your brainwaves into colored light

TIME REQUIRED: A WEEKEND
COST: $100–$130

MATERIALS

- » **Beanie, double-layer knit** with a wide, fold-over flap
- » **Copper foil sheet, thin**
- » **Conductive fabric, non-woven**
- » **Foam, high-density**
- » **Copper tape**
- » **Weatherstripping adhesive tape, 1½" wide**
- » **RG174 shielded wire for electrodes**
- » **ECG snap lead**
- » **Electrode, small adhesive, one-time-use, with snap** such as SparkFun SEN-12969
- » **NeuroSky ThinkGear ASIC Module PCB** can be hacked from Mindflex headset
- » **Flora Neopixel RGB LEDs (4)**
- » **TinyLily Arduino-compatible microcontroller**
- » **TinyCircuits mini FTDI USB adapter**
- » **Heat-shrink tubing, colored**
- » **Metal and fabric glue, all-purpose clear**
- » **Cardboard**
- » **Yarn, thick white** for pom-pom (find tutorials online)
- » **Felt or microfiber fabric**
- » **Thread**
- » **Battery extension with JST connector**
- » **Battery, lithium ion, 3.7V 500mAh**
- » **Switch, tactile pushbutton with 2 leads**
- » **Connectors, 4-pin: male (1) and female (1)**
- » **Wire, insulated, 0.5mm** flexible but sturdy
- » **Kapton or electrical tape**

TOOLS

- » **Soldering iron and solder**
- » **Wire stripper**
- » **Scissors**
- » **Sewing needle**

EVER WANTED TO VISUALIZE YOUR BRAIN ACTIVITY? Electroencephalography (EEG) uses electrodes placed against the scalp to detect the tiny electrical changes that occur when neurons fire. By amplifying these signals through a computer, you can observe brain activity in real-time.

Using an off-the-shelf EEG module, you can make a beanie with lights that change color and intensity in response to the wearer's level of attention and relaxation. You can even use the hat to control interfaces on your computer! The flap and double-layer knit hide the electronics, so it looks and feels just like a comfortable cap.

HOW IT WORKS

A NeuroSky ThinkGear ASIC Module (TGAM) circuit board is connected to a TinyLily Arduino-compatible microcontroller (see page 35) running a program that translates brain activity into light through four NeoPixel RGB LEDs. No modifications are made to the TGAM software or hardware — rather, you integrate the ThinkGear chip into the circuit via a simple mechanical adaptation that allows you to connect it to the board.

The TinyLily is easily accessible via a micro USB connector hidden inside the hat's pom-pom. This lets you create interactive "brain art" on your computer, using a programming language such as Processing. Simply connect to your computer via USB (or wirelessly if you make the hat with Bluetooth), and import the real-time serial stream into Processing. Then customize the Processing sketch so that levels of "attention" and "meditation" will control variables such as color, opacity, movement, coordinates, and rotation.

UNDERSTANDING ELECTRODES

An EEG device has two types of electrodes: the EEG electrode, which records brain signal from the scalp, and the reference electrode, which tells the EEG device what "no brain signal" looks like (i.e. baseline). I built the hat with a dry EEG electrode — constructed from copper foil, conductive fabric, foam, and conductive tape — which provides a decent signal quality and is less messy than traditional gel EEG electrodes. You'll need to ensure constant contact of this electrode with your forehead, and to keep the area sweat free.

MAKE YOUR OWN

This project requires soldering and some sewing. Detailed step-by-step instructions can be found at makezine.com/go/eeg-beanie.

GET GLOWING!

Put on the hat, placing the reference electrode over the mastoid bone behind your ear. Snap the sensor wire to the electrode and switch the hat on. The lights in the pom-pom should begin to flicker as the ThinkGear chip starts processing your brain activity. Take slow, deep breaths to relax and watch as the color changes. Then try focusing on a difficult mental math exercise and watch as the pom-pom glows the opposite hue! ●

IO FLAMENT
is a neuroscience student with a passion for art, music, and tinkering with DIY electronics. illumino.io

Get complete step-by-step instructions and photos at makezine.com/go/eeg-beanie.

Io Flament

GET YOUR GIF ON

Download animated GIFs to a 16×16 LED matrix and wear your favorites **WRITTEN BY ROBERT TU**

Hep Svadja

WANT TO ROCK BIG ANIMATIONS ON YOUR LED WEARABLES? There's an easy way to download animated GIFs onto a 16×16 RGB LED matrix using Processing software, Arduino, and the Teensy microcontroller. We designed this project for our new MeU Square matrix but you can use it with any home-built NeoPixel LED grid. We've made it easy — all the required code and libraries are on our Github page at github.com/MeULEDs.

1. FIND A 16×16 ANIMATED GIF

Search Google Images for a subject you like, then use the Search Tools to narrow it down. First select Size→Exactly and enter 16 for height and width. Then select Type→Animated.

Or you can scale any existing animated GIF to 16×16 in image-editing software.

2. TURN THE GIF INTO AN ARDUINO PROGRAM

Open our Processing sketch *MeU_Square_GIF_Converter_Teensy.pde*. It automatically turns your GIF into an Arduino program (also called a sketch). Just select your GIF, then choose a filename and directory for your Arduino sketch.

3. CONFIGURE YOUR ARDUINO SKETCH

Open your Arduino sketch in the Arduino IDE. You can adjust the GIF playback speed by changing the Metro AnimateTimer value; usually 70 to 100 is normal. The smaller the number, the faster the animation moves.

You can also adjust the setBrightness value but *do not exceed 40!* These LEDs suck up a lot of power, and a 16×16 displaying white at full brightness will draw up to 15 amps. Unless your matrix is plugged into the wall and is designed for 15A of current draw, do not go past 40 or you'll do some serious damage.

4. UPLOAD THE SKETCH TO THE TEENSY

Connect the Teensy to your computer's USB port, then press the Upload button in Arduino. If Teensyduino is installed correctly, it should pop up and upload the code. That's it. Your matrix has got its GIF on! ◗

TIME REQUIRED:
30–90 MINUTES
COST:
$150–$250

MATERIALS

» **MeU Square LED matrix** A new open-source wearable 16×16 RGB LED display you can put into clothing, shipping in 2015. themeu.net

—OR—

» **Teensy 3.1 microcontroller Arduino compatible, USB-based,** $20 from pjrc.com/teensy

» **RGB LEDs, WS2812B type (256)** such as NeoPixels, Adafruit #1655. You'll have to solder or connect the LEDs yourself in a zig-zag configuration. You can also use 60-LED strips (Adafruit #1138), or daisy-chain four 8×8 matrixes (Adafruit #1487). Read the NeoPixel Über-guide at learn.adafruit.com/adafruit-neopixel-uberguide/overview

» **Battery, lithium ion, 2,000mAH** such as Spark-Fun #8483. You can also use a power supply, but be aware of the brightness setting of the LEDs (see Step 3).

TOOLS

Computer with the following software:

» **Arduino IDE 1.0.5 or 1.0.6** free from arduino.cc/downloads.

» **Arduino libraries** from github.com/MeULEDs: *Adafruit-GFX-Library, Adafruit_NeoPixel, Adafruit_NeoMatrix, SimpleTimer,* and *Metro-Arduino-Wiring*

» **Teensyduino** free from pjrc.com/teensy/teensyduino.html. This add-on is required to upload code to the Teensy using the Arduino IDE.

» **Processing** free from processing.org

» **Processing libraries** from the Github page: *controlp5* and *gif-animation*

» **Project code:** *MeU_Square_GIF_Converter_Teensy.pde* from the Github page

ROBERT TU

is the founder of MeU, a company developing interactive and reactive clothing. He's an electrical engineer and designer, IBM alumnus, and graduate of University of Waterloo and OCAD University in Toronto.

For complete step-by-step instructions and video, visit makezine.com/downloading-animated-gifs.

Shishi-Odoshi Fountain

This Japanese water feature literally rocks — to scare hungry critters away from your garden

written by Andrew Terranova

Time Required:
A Weekend
Cost:
$50–$150

Materials

» **Bamboo, about 8'–20' total length, diameters from 1"–3"** see Steps 1–2
» **Submersible pump, low volume** 80 gallons per hour (GPH) or less
» **Basin, pond, or sealed planter**
» **Clear vinyl tubing, ½", 3'–10' length**
» **Steel rod, zinc plated, ³/₁₆" diameter, 2'–3' length**
» **Rope, ¼", about 6' length**
» **PVC pipe, ½", about 1' length**

Tools

» **Handsaw**
» **Drill and drill bits** I recommend using a drill press.
» **Forstner bits or hole saws**
» **Bench vise**
» **Hacksaw**
» **Hammer**
» **Pliers**
» **File**
» **Ruler**
» **Level**
» **Pencil or fine-tip marker**

IN JAPANESE, *SHISHI-ODOSHI* **MEANS "DEER SCARER." THIS TYPE OF FOUNTAIN SLOWLY FILLS WITH WATER, AND THEN SUDDENLY TIPS** — making a gentle knocking sound that will chase away any critters eating your garden. Many Zen gardens also use these rocking fountains as a meditation aid. Here's how to build your own.

1. SELECT YOUR BAMBOO

Each node is a barrier inside the bamboo. You'll want one upright piece to have few or no nodes, as you'll need to run tubing through it. For the rocking piece, you want a node in the middle, which will form the bottom of the scoop that fills with water.

> **NOTE:** Working with bamboo is not like working with dimensional lumber. Bamboo surfaces are irregular, and size and shape change along the length. Be prepared to adjust instructions accordingly.

2. DETERMINE DIMENSIONS

I built a small fountain to fit in a planter just 8" across, and used 6 or 8 linear feet of bamboo. Marty Marfin in the *Make:* Labs built one 3' tall (shown here) and used about 20' total.

Whatever size you design, you'll need a large-diameter piece for the top beam, 2 upright pieces that'll fit into this beam, and a smaller-diameter piece for the spout, which also fits into the beam.

Choose another fairly large-diameter piece for the water scoop.

3. CUT AND DRILL BAMBOO

Cut your beam and uprights to length. Measure the tops of the uprights carefully and drill holes in the beam to accept them (Figure Ⓐ).

Drill a third hole centered in the front of the beam, sized to accept the spout. But leave the spout piece extra long, as you'll figure out the final length after some testing.

Test-fit the uprights, then drill a ¾" hole near the bottom of one of them, for routing the ½" tubing. I used Forstner bits; they leave beautifully clean holes (Figure Ⓑ).

Cut the water scoop piece so it's got about the same length on either side of a node.

4. CUT STEEL RODS

Use a hacksaw to cut a length of ³⁄₁₆" rod to span your uprights, plus a few inches on either side. This will be the axis of the water scoop.

Cut a shorter rod to fit between your uprights without touching; you'll use this to test the pouring action (Figure Ⓒ).

5. MAKE THE WATER SCOOP

Drill a ⁷⁄₃₂" hole just behind the central node and straight through the center of the bamboo.

Insert the shorter, test axis through the hole, and clamp it in a bench vise so the axis is straight up. Now trim one end of the bamboo at a shallow angle of about 30°, keeping the saw perpendicular to the ground (Figure Ⓓ). Just cut off a little; you may have to trim more later.

6. THREAD THE TUBING

Feed the ½" tubing up through the upright you drilled (Figure Ⓔ), into the top beam,

> **TIP:** Drill with an extra-long bit — or just hammer a piece of rebar — to punch through any nodes blocking your path.

F

G

H

I

and out the spout. You can use a piece of rod or another tool to help guide the tubing.

7. TEST THE POURING ACTION
Determining exactly where to mount the water scoop on the uprights is key. Dry-fit the bamboo frame together and use a temporary crosspiece of wood or bamboo to help support it.

Trim the tubing if needed, and connect it to the pump. My pump came with a valve that can partially restrict the water flow. You want it to pump as slowly as possible.

Run the pump and adjust the height of the water scoop between the 2 uprights to see how it works (Figure H). You may need to adjust the length of the spout and/or the scoop so that water can pour from one to the other.

> **CAUTION:** I recommend testing the pump outside — it'll get splashy.

8. MARK AND DRILL
Once you've found the right placement for your water scoop, mark the uprights where the test axis is aligned.

Disassemble the frame and tubing, and drill ³⁄₁₆" holes at your marks (Figure F). Reassemble.

9. INSTALL THE WATER SCOOP
Push the axis rod through the first upright, tapping gently with a hammer if needed. Cut 2 short pieces of small-diameter bamboo. You'll use them as spacers to keep the scoop from moving too far from side to

side. Place one spacer on the rod, then the scoop, then the second spacer (Figure G).

Now push the rod all the way through the far side of the second upright. If you wish, cover the exposed ends of the rod with 2 more short pieces of bamboo to match the spacers.

10. MAKE THE KNOCKER
In some traditional fountains, the water scoop tips down and strikes a rock or a basin to create the deer-scaring noise.

This one uses a lower crosspiece lashed to the frame to provide the desired knocking sound when the scoop tips back up. Test for desired location, then use thin rope to lash it on (Figure I).

11. TROUBLESHOOT AND ADJUST
Check the flow of water and the rocking action. Adjust the spout angle if needed. I found I needed to straighten the vinyl tubing inside the spout by inserting a short collar of ½" PVC pipe to better direct the flow. Your mileage may vary.

If the water scoop doesn't tip and dump after it fills, the back of the scoop weighs too much. You may have to saw some off to adjust the balance.

If it spills but then doesn't flop back into position, the front end is too heavy. Trim the front or add some weight inside the back end.

Now put your rocking fountain in your garden to make a space that's peaceful for you — and for your plants. ◓

+ Thanks to Bamboo Sourcery (bamboosourcery.com) in Sebastopol, California, for the bamboo and photo location.

> Get more photos and tips, and share your build at makezine.com/japanese-water-fountain.

ANDREW TERRANOVA is an electrical engineer, writer, and electronics and robotics hobbyist. He's an active member of the Let's Make Robots community and has taught robotics in primary school and created robotics exhibits for the Children's Museum of Somerset County, New Jersey.

Written by Enrique DePola

DIY
"Magic Shell"
Chocolate Dip

Magically hardens in seconds — easy to make in minutes!

Time Required:
5 Minutes
Cost:
$5–$10

Ingredients

» ½ lb (250g) dark chocolate, 66% to 72% cacao, chopped
» 1 cup (200g) refined coconut oil not virgin or unrefined
» 6 Tbsp (125g) honey or light (clear) corn syrup

THIS DELICIOUSLY SNAPPY, YET FUDGY CHOCOLATE DIP IS SURPRISINGLY SIMPLE TO MAKE, needing just chocolate, coconut oil, and sweetener. The secret is the saturated fat in the coconut oil, which hardens when chilled.

We liked Max Falkowitz's recipe (makezine.com/go/magic-shell) using gourmet bittersweet chocolate balanced with a touch of corn syrup. So we made our own version with honey as the sweetener, and it didn't disappoint.

1. Combine ingredients in a ceramic or glass bowl.

2. Microwave in 15-second bursts, stirring well in between, until completely melted.

3. Store at room temperature and stir if it separates.

4. Drizzle it on ice cream — or dip ice cream bars or cones right in it.

5. Let harden about 30 seconds, or until the glossy appearance takes on a matte finish. Crunch! ◗

See more photos and share your recipe and tips at makezine.com/projects/diy-magic-shell.

ENRIQUE DePOLA is a ceramics artist, skater, and amateur chef. Formerly a *Make:* engineering intern in Sebastopol, California, he is on the road again.

Hep Svadja

Dirty Dish Detector

Combine a webcam with real-time computer vision software to alert you when dishes pile up.

Written by Nick Normal and Jason Kridner

Time Required:
A Weekend
Cost:
$140–$170

JASON KRIDNER is a veteran of Texas Instruments and co-founder of the nonprofit BeagleBoard.org Foundation, promoting the design and use of open-source software and hardware in embedded computing.

NICK NORMAL is a Queens-based artist and maker, and a lifelong biblioholic. He is a former five-year resident of Flux Factory, co-organizer for World Maker Faire (NYC), and advocate for all things geekathon.

IS YOUR SINK FULL OF DIRTY DISHES LEFT BY UNKNOWN SHIRKERS? Do you require better discipline around your home or hackerspace to stay on top of the kitchen chores?

The Dirty Dish Detector combines an affordable BeagleBone Black single-board computer and a Logitech webcam — along with plenty of open-source software — to alert you the instant dishes get left in the sink. The key is free computer-vision software called OpenCV, which can recognize shapes like a sink or a bowl and figure out when something's changed.

You can find step-by-step instructions on the project page online at makezine.com/dirty-dish-detector. Here's how it works:

1. CONFIGURE THE BEAGLEBONE BLACK

Update the BeagleBone Black's MicroSD card to run the latest Debian operating system, which includes OpenCV and Python libraries (Figure Ⓐ).

Then, configure it to automatically connect with your wi-fi network.

2. 3D PRINT AN ENCLOSURE (OPTIONAL)

You could use a plain project box but we 3D-printed a BeagleBone Black enclosure by Logic Supply that also accommodates the USB hub (with a little drilling) and the webcam. The 3D files are free on Github (Figure Ⓔ).

> **NOTE:** This project assumes some familiarity with the command line, along with the ability to administer your local area network. A little experience with coding (be it Python, PHP, or even HTML) will go far in this project. All of the necessary code is supplied and documented on the project page online, but you'll want to customize it to your liking.

Dishes are clean!

Materials

» **Getting Started with BeagleBone Black Kit** Maker Shed #MSGSBBK2
» **Webcam, widescreen** such as the Logitech C270
» **MicroSD card**
» **AC adapter with USB, 5V/3.6A**
» **AC adapter plug, M-type**
» **USB hub, 4-port**
» **USB wi-fi dongle,** Netgear G54/N150

Tools

» **Computer with USB port**
» **Hot glue gun**
» **Nippy cutter**
» **3D printer (optional)**
» **Zip ties**
» **Hardware, various** for mounting the project

3. INSTALL THE DETECTOR ABOVE YOUR SINK

How you mount the Dirty Dish Detector is up to you. Some craftiness will be necessary, depending on whether you have cabinets above your sink, what they're made of, and how many sinks you need to detect (Figure B).

Place the Dirty Dish Detector at an adequate distance to cover as much of your sink or sinks as possible. Experiment with the webcam's lens and angle before you install your Detector.

4. RUN CLOUD9 AND PROGRAM THE BOARD

The BeagleBone automatically connects to your wi-fi network. Use the Cloud9 IDE to interface with your Detector and program it, using the ready-made Python scripts we provide.

5. CAPTURE CLEAN AND DIRTY SINK IMAGES

Now you'll "train" the OpenCV software to understand what it's seeing. First, test the camera by running the script *camera-test.py*. Then run the *sink-empty.py* script to take a picture of your empty sink (Figure C). This gives OpenCV a reference image to compare against when it processes photos, looking for culprit cups and dirty dishes.

Then throw some dishes in and take another picture by running the *sink-latest.py* script (Figure D).

NOTE: Good lighting is important for reliable detection. Turn on any available kitchen lights and experiment.

6. SET UP NOTIFICATIONS AND AUTOMATION

Once the system knows what to compare against — a clean sink — you can program it to send an email and/or MMS when an unclean sink is detected. In fact, these notifications will happen with every "status change" in the sink, so you'll be notified not only when the sink goes from "clean" to "dirty," but also vice versa.

Lastly, go back out to the command line and set up a *crontab* to automate the system to take a photo every 5 minutes. The webcam's status LED will light up whenever the script runs.

Now you've got a tireless partner in your fight against unwashed dishes!

Do you have another idea for a robust, self-contained Dirty Dish Detector? Perhaps you designed an enclosure to blend into your kitchen environment? How about a Dirty Dish *Detective* that also snaps a photo of the perpetrator and tweets it for public shaming! Share your ideas on the project page online.

For complete step-by-step instructions, images, code, and links, visit makezine.com/dirty-dish-detector.

Maker HANGAR

Maker Trainer R/C Airplanes

Learn to build — and fly — your own radio-controlled airplanes with the Maker Hangar video tutorial series.

Written by Lucas Weakley

LUCAS WEAKLEY is studying aeronautics engineering at Embry Riddle Aeronautical University. He also makes and sells aircraft kits at lucasweakley.com. He's a certified AutoCAD draftsman, an Eagle Scout, and the host of *Make:*'s Maker Hangar video series at makezine.com/makerhangar.

I THINK WE'VE ALL BEEN FASCINATED BY FLIGHT AT ONE POINT IN OUR LIVES. Whether that fascination leads to folding paper airplanes or piloting full-sized aircraft, we all dream to make something fly. And many of us get there using radio-controlled (R/C) hobby aircraft.

Plenty of toys can give you limited control of a flying craft, but to get the full sense of flight you need to dive right into the R/C community. You'll soon be doing exciting activities like acrobatics, speed trials, formation flying, combat, slope soaring, and aerial photography.

However, it can be daunting to get started with your first R/C plane. What motor and speed controller should you get? How should you charge the batteries? What is a BEC and why do you need one? How do you fly?!

To answer all these questions *Make:* created Maker Hangar, a one-stop, free resource that anyone can use to easily get into the R/C hobby. Maker Hangar consists of 23 video tutorials, three aircraft you can build, and a community of more than 1,000 members all sharing pictures, videos, and knowledge. Join us at makezine.com/makerhangar.

LEARN TO BUILD AND FLY — ALL FROM VIDEOS

The Maker Hangar video series covers the basics of electronic R/C aircraft components, then shows how to build, set up, and fly your first trainer airplane. Following that you'll build a tricopter that's great for aerial video, FPV (first-person view), and just fun flying — find it in the next *Make:* magazine — and a smaller, tougher trainer plane to hone your flying skills.

Turn the page for an up-close look at the Maker Trainer planes and get building!

Kent Weakley

The Maker Trainer 2 is a tough, capable, and fun-flying R/C plane you can build and fly by following along with the Maker Hangar video series.

MAKER HANGAR: THE EPISODE GUIDE

Maker HANGAR

Maker Trainer
Super stable for beginners, jumbo sized for payloads

THE ORIGINAL MAKER TRAINER USES ALL STANDARD ELECTRONIC R/C PARTS AND WAS DESIGNED WITH A LARGE FUSELAGE TO MAKE IT EASY TO WORK INSIDE. Flight characteristics were taken very seriously too — the Maker Trainer has almost a 5-foot wingspan, so it can glide for long distances and carry plenty of weight. Because the plane is big and heavy, it's also more stable and can fly in mild winds; although if you're a beginner it's preferable to fly in no winds until your reaction times decrease and your confidence grows.

This plane is also very crash-resistant: It's got a double-layered fuselage and a wooden spar running the length of the wing, and its propeller is mounted behind the fuselage to protect it in the event of a crash.

Another big consideration with this big plane: how to transport it to the flying field. For portability, the wings fold back parallel to the fuselage. This drastically decreases the dimensions of the airplane so you can fit it in the back seat of almost any car.

All in all, the original Maker Trainer is a great first plane if you want a very stable and durable trainer and don't mind spending a little more time and money to build it.

Watch the original Maker Trainer how-to video series at makezine.com/makerhangar.

Time Required:
A Weekend
Cost:
$280–$300

Materials
Airframe:
» Foam board, RL Adams Readi-Board, ³⁄₁₆"×30"×20" (4)
» Plywood, ⅛"×2"×2" for motor mount
» Popsicle sticks (3)
» Socket cap screws, M3×16mm (4)
» Lock nuts, M3 (4)
» Push rods, 36"×0.047" (2)
» Packing tape, heavy duty aka carton sealing tape
» Wood square dowels: ³⁄₈"×³⁄₈"×36" (2) and ½"×½"×36" (1)

Electronics:
» Motor, brushless outrunner, 1,800kV, NTM Prop Drive 28-36
» Motor hardware, NTM Prop Drive 28 Series
» Propeller, 6"×4", APC
» ESC, 30A, Turnigy Plush
» LiPo batteries, 2,200mAh 3S (2)
» Servomotors, 9g Micro (3 or 4) Get 4 servos if you want rudder control.
» Servo extensions, 12" (2)
» R/C transmitter, 4+ channels
» R/C receiver, 4+ channels

Lucas Weakley

MORE GREAT R/C PROJECTS AND KITS

BROOKLYN AERODROME FLYING WING
Fly in tight spaces, handle turbulence, even carry a camera with this highly maneuverable flier! Build it from scratch at makezine.com/the-towel or get our kit, item #MSFW1 from the Maker Shed, makershed.com.

R/C OMNIWHEEL ROBOT
Our new Make: Motor Shield (#MSMOT02, makershed.com) lets you drive all kinds of projects using your standard R/C gear! Build this easy holonomic "Kiwi drive" robot platform that moves instantly in any direction, at makezine.com/kiwi.

3D ROBOTICS QUADCOPTERS
These autonomous choppers are great for aerial photography or honing your R/C flying skills. The 3DR Quad Kit (#MK3DR01, makershed.com) is a DIY build with the APM autopilot; the fully assembled Iris (#MK3DR03) uses the programmable Pixhawk system.

Maker Trainer 2

Our most durable, compact, quick to build, and fun to fly plane

Time Required:
An Afternoon
Cost:
$160–$180

Materials

Airframe:
» Foam board, RL Adams Readi-Board, ³⁄₁₆"×30"×20" (2)
» Maker Trainer 2 parts kit $15 from lucasweakley.com/product/maker-trainer-2-kit, includes:
  » 3D-printed motor mount
  » 3D-printed control horns (6)
  » Socket cap screws, M2.5×8mm (4)
  » Hex nuts, M2.5 (4)
  » Push rods: 4"×0.047" (2); 3"×0.047" (2); 2.5"×0.047" (1)
  » Push rod connectors (5)

Electronics:
» Motor, brushless outrunner, 1,400kV, HURC 300 "Blue Wonder"〉
» Propeller, 8"×3.8", APC Slow Fly
» Electronic speed controller (ESC), 18A
» LiPo batteries, 1,300mAh 3S (2)
» Servomotors, 9g Micro (3) Get 5 servos if you want rudder control.
» Servo extensions, 18" (2) Get 3 if you're installing rudder servos.
» R/C transmitter, 4+ channels
» R/C receiver, 4+ channels

Tools
» Straightedge
» Cutting board
» Screwdriver set
» Scissors
» Spray adhesive
» Hobby knife such as an X-Acto
» Packing tape, heavy duty
» Sandpaper
» Drill and drill bits
» Velcro
» Hot glue gun
» Cyanoacrylate (CA) glue aka super glue
» Cutters
» Pliers
» Soldering iron and solder

WE DESIGNED THE NEWEST MAKER TRAINER TO BE SMALLER, EVEN MORE RUGGED, AND EASIER TO FLY THAN THE ORIGINAL. The motor is still safely tucked in back, but the MT2 is half the size and weight of the MT1, so it doesn't need to be folded and it's more rigid and resilient to crashes. No wood supports are needed, only two sheets of foam board. Build it in just a few hours.

The plane has amazing flight characteristics too. With its KFm2-type stepped airfoil, the MT2 is completely resistant to stalls and has a flight speed envelope from walking speeds to high-speed full-throttle passes. It's also capable of mild acrobatics such as loops, rolls, and flat spins. Naturally, it does get thrown around by the wind a little more than its larger brother.

1. LAY OUT THE PLANS
Download the PDF plans, print them out, and tape them to the ¼" foam board.

2. CUT OUT THE FOAM
Cut the pieces, then score the foam to free up all the control surfaces.

3. ASSEMBLE THE WING
Use spray adhesive to glue the 2 parts of the KFm2 airfoil, lining up the wingtips and leading edge. Use sandpaper to taper the leading edge.

4. TAIL BOOMS AND FUSELAGE
Fold and glue the fuselage and tail booms.

5. GLUE THE PLANE TOGETHER
Glue the tail booms to the wing, the elevator to the booms, the vertical stabilizers to the booms, and finally the fuselage to the wings.

6. PAINT (OPTIONAL)
Waterproof the plane with polyurethane, and then tape off any design and spray-paint.

7. INSTALL THE ELECTRONICS
Mount the motor and glue the servos in place.

8. CONNECT CONTROL SURFACES
Mount the control horns in the plane's control surfaces, then link them to the servos.

9. PLUG IN ALL THE WIRES
Plug the ESC and servos into their corresponding ports in the R/C receiver.

10. PROGRAM THE RADIO
Make sure all control surfaces are moving the right way and at the proper proportions.

11. FLY!
Tweak the center of gravity until it flies without dipping, and trim the plane to make it fly level.

The Maker Trainer 2 is the perfect fun-flying, durable, cool-looking trainer for anyone wanting to learn to fly R/C airplanes. Build one! ◐

Watch the complete Maker Trainer 2 how-to video series at makezine.com/makerhangar.

Smart Rat Trap

Arduino + desperation = 21st-century rodent control.

Written by Isa McKenty, Immanuel McKenty, Adam McKenty, and Eli McKenty

WE LIVE ON A SMALL ISLAND WITH A LARGE POPULATION OF RATS.
Since we're opposed to capital punishment of uninvited guests, even
furry ones, we try to catch them with live traps when they sneak inside.
After years of failed attempts with store-bought traps, we realized we
had to make a trap as clever as the rats.

Rather than an enclosed cage, which they seem to avoid, we used
a suspended basket that leaves an unobstructed view at rat eye level.
The cage is suspended by two neodymium magnets. When an infrared
proximity sensor detects a rat below, it triggers two electromagnets that
repel the suspension magnets, dropping the cage over the rat.

Over several years (and rats), the trap has gradually improved. The
sensitivity is recalibrated every 10 minutes in software running on an
Arduino. An RGB LED indicates trap status, a buzzer sounds the alarm
when something gets caught, and a handle makes it easy to carry.

1. BUILD THE CAGE

1A. MARK PLYWOOD
Mark out your plywood as shown in the cutting diagram (Figure **A**).

1B. DRILL HOLES
Using hole saws or Forstner bits, drill the two 1⁵⁄₁₆" holes in the base
(Figure **B**) and the two 1⁵⁄₈" holes intersecting the cage rim.

Drill two ⁵⁄₁₆" holes in the middle of the cage top, the same distance
apart as the emitter and receiver on your sensor (about ¾"). Then drill
two ¼" holes in the top, 5" apart and in line with the sensor holes; these
are for mounting the disc magnets.

1C. CUT THE PLYWOOD
Using a jigsaw, cut out the base, rim, and top (Figures **C** and **D**). (Drill
a hole first to get the blade started.) Touch up with 120-grit sandpaper.

1D. ASSEMBLE THE CAGE
Wrap a 7"×58" strip of metal mesh around the cage top and staple in
place (Figure **E**). Staple the other end of the mesh cylinder into the
cage rim (Figure **F**). If your rodents are voracious chewers, staple a
small piece of screen over the sensor holes inside the cage.

1E. PAINT (OPTIONAL)
Give the base and cage a coat of paint or two to make it look snappy.

IMMANUEL, ADAM, ISA, AND ELI MCKENTY are four autodidact brothers who enjoy playing music, writing, and tinkering with computers, electronics, and themselves. They hail from Cortes Island on Canada's west coast, where the weather is wet and rodents are plentiful. They occasionally post the results of their experiments at autodidacts.io.

**Time Required:
Half a Day
Cost:
$50–$100**

Materials

» **Arduino Uno R3 micro-controller board** Maker Shed #MKSP99, makershed.com
» **Mini solderless breadboard** Maker Shed #MKKN1-B or MSBR1
» **Breadboard jumper wires (8)** Maker Shed #MKSEEED3
» **AC/DC adapter, 12V, 1A, center positive**
» **Capacitor, 4,700µF**
» **Resistor, 1kΩ**
» **Resistors (optional): 100Ω (1), 50Ω (1)**
» **Transistor, NPN, 2N3904**
» **LED, RGB common cathode (optional)**
» **Speaker, 0.5W 32Ω (optional)**
» **Relay, SPDT, 5V DC, type HRS4**
» **Silicon power rectifier diodes, 1N4001 (2)**
» **Infrared proximity sensor** Sharp GP2Y0A21YK
» **Neodymium magnets, ½"×⅛" disc (2)**
» **Magnet wire, 30 AWG, 70'**
» **Pan head screws, #8 (6)**
» **Hex bolts, ¼"×1" (2)**
» **Hex nuts, ¼" (2)**
» **Threaded rods, ⁵⁄₁₆"×6" (2)**
» **Nuts, ⁵⁄₁₆" (10)**
» **PVC pipe, 1" ID, 5' length**
» **PVC pipe fittings, 90° elbows (2)**
» **Plywood, ⁵⁄₈"×4'×2'**
» **Plywood, ⅜"×6"×5½"**
» **Metal mesh, ¼", 7"×58"** aka hardware cloth
» **Electrical tape and heat-shrink tubing**
» **Masking tape**
» **Paint** (optional)

Tools

» **Soldering iron and solder**
» **Stapler**
» **Side cutters**
» **Jigsaw**
» **Drill and bits**
» **Hole saws or Forstner bits**
» **Hot glue gun**
» **Lighter or mini torch**
» **Sandpaper**
» **Magnetic compass or compass app on your mobile device**
» **Computer running Arduino IDE** free download from arduino.cc/en/Main/Software
» **Project code** download the free Arduino sketch from github.com/Photosynthesis/make-smart-rat-trap

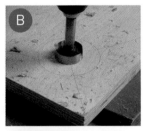

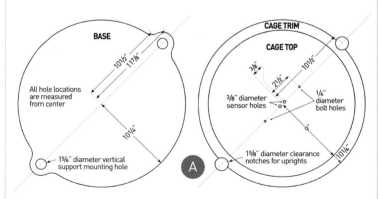

BASE

10½"
11⅞"

All hole locations are measured from center

10¼"

1⁵⁄₁₆" diameter vertical support mounting hole

CAGE TRIM

CAGE TOP

3⁄8"
2½"
10½"

⅜" diameter sensor holes

¼" diameter bolt holes

10¼"

1⅝" diameter clearance notches for uprights

A

TIP: To draw perfect circles on plywood, put a small nail in the center
and use a stiff wire with a crook at each end to keep your pencil in
place as you draw the circle.

G

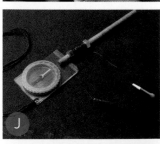

H

I

J

K

2. BUILD THE SUPPORT FRAME
2A. CUT THE PVC PIPE
Cut your PVC pipe into 4 sections: a 19⅝" crossbar, two 12¼" uprights, and a 6" handle.

2B. DRILL THE PVC FRAME
Drill two 5/16" holes 5" apart through the handle. Drill a matching pair centered in the crossbar.

2C. ASSEMBLE THE FRAME
Use 2 PVC elbows to attach the crossbar to the uprights, making sure the bolt holes are parallel with the uprights. Drill pilot holes and fasten the elbows with ¾" #8 pan head screws. (You could use cement, but screws make it removable.)

3. MAKE THE ELECTROMAGNETS
3A. WIND THE COILS
Wind 2 nuts onto each threaded rod, 1" apart, 11/16" from one end. Wrap electrical tape over the threads between nuts (Figure G).

Measure out 35' of magnet wire. Tape it to the threaded rod between the nuts, leaving 6" of extra wire (wrap this around the bolt shaft out of the way for now).

Chuck the threaded rod in your drill, and cover the exposed threads with masking tape so they won't damage the wire's insulative coating. Now carefully wind the wire on between the 2 nuts (Figures H and I). Secure it with a layer of tape, again leaving 6" of wire hanging out.

Repeat for the other electromagnet, making sure the direction of winding is the same.

3B. TEST POLARITY
Hook each electromagnet up to a 12V power source briefly, with the lead from the outside of the coil connected to positive (+). Use a magnetic compass (or smartphone compass app) as shown to determine the magnet's polarity (Figure J). Record which end of the compass needle (south or north) points toward the active electromagnet.

4. CONNECT THE ELECTRONICS
4A. MAKE THE ELECTRONICS PLATFORM
Cut a 6"×5½" rectangle of ⅜" plywood. Drill a pair of 5/16" holes through it, 5" apart.

Use small beads of hot glue to attach the Arduino, capacitor, and 5V relay onto the platform board. If you're using a speaker, glue that on too (Figure K).

4B. WIRE THE BREADBOARD
Stick a mini breadboard onto the platform next to the Arduino. Wire the transistor as shown in the wiring diagram, with the 1kΩ current-limiting resistor between its base and the Arduino pin D12 (Figure L).

For audio and visual feedback of trap status, add the RGB LED and speaker as shown. Holding the RGB LED with the longest leg second from the left, the pinout from left to right is Red, Common Cathode, Green, Blue. Solder a 100Ω resistor to the common cathode and a 50Ω resistor on the red leg. Put a small piece of heat-shrink tubing over each leg (including the resistors. Then shrink a larger piece over them all (Figure M). Connect the cathode to ground, the red leg to D7, green to D6, and blue to D5 on the Arduino.

Connect the speaker between GND and D8.

4C. SOLDER THE DIODE AND CAPACITOR
Solder a silicon power diode between the capacitor's negative (–) terminal and the relay's common (COM) terminal; the diode's cathode (marked with a line around the diode) should be connected to the relay (Figure N).

4D. SOLDER THE RELAY
Run a wire from the capacitor's positive (+) terminal to the relay's normally open (NO) terminal. Solder another wire onto the same relay terminal and plug it into the Arduino's Vin pin.

Solder a silicon power diode across the relay's coil. Solder a wire to the relay pin where the diode's anode is attached, and connect the other end to the transistor's collector at the breadboard. Run a wire from the other relay coil terminal (cathode side of the diode) to the Arduino's +5V pin. Strip this end a little extra long so you can solder another wire to it (Figure O).

4E. WIRE THE SENSOR
Your sensor should have a 3-pin female plug on the end of its lead. Cut 3 pieces of appropriately colored hookup wire, and use these to extend your sensor's leads. Plug the black lead into the Arduino's GND, and the yellow lead (signal) into A1. Solder the sensor's power wire to the relay power wire at the Arduino's +5V pin (Figure P).

Glue the sensor onto the bottom of the electronics platform, making sure the infrared emitter and receiver match the 2 holes drilled in the top of the cage (Figure O).

4F. INSTALL THE ELECTROMAGNETS
Insert the 2 threaded rods into the holes in the electronics platform and secure them with nuts underneath.

Cut the electromagnet leads to length. Use a mini torch, lighter, or sandpaper to remove the shellac from the ends of the leads.

Solder the leads from the outside of both electromagnets' windings to the relay's common (COM) pin.

Solder the wires from the inside of the coils onto the capacitor's negative (–) wire at the Arduino's GND pin.

5. FINAL ASSEMBLY

5A. ATTACH THE SUPPORT FRAME
Insert the PVC frame into the holes in the base (Figure ❓). From the underside of the base, drill 2 pilot holes from the inside of each pipe, then secure the pipe with #8 pan head screws.

5B. ATTACH ELECTRONICS AND HANDLE
Slide the electronics assembly's threaded rods up through the holes in the horizontal crossbar. Thread down 2 nuts to clamp it in place. Use 4 more nuts to attach the handle on top (Figure ❓).

5C. INSTALL THE PERMANENT MAGNETS
Thread two ¼"×1" bolts through the holes in the top of the cage. Secure with 2 nuts on the bottom.

Test the polarity of 2 neodymium magnets, then snap them onto the bolt heads, oriented so they'll repel the electromagnets (Figure ❓). For example, if the electromagnets' north pole is down, then the permanent magnets' north pole should face up — like repels like.

Finally, squeeze the cage into place between the support posts.

5D. PROGRAM THE ARDUINO
Download the Arduino sketch from Github and upload it to the board.

5E. TEST!
Slide the cage up until it snaps onto the magnetic supports (Figure ❓), then plug your AC/DC adapter into the Arduino's power jack. The LED will blink green while the Arduino calibrates the sensor's threshold, then turn solid green once it's set and ready to trigger.

Test the trap by slipping a rodent-sized object under the center of the cage. The LED will flash red and the Arduino will close the relay with a click to fire the electromagnets, dropping the cage. A blue LED means the trap has been triggered and may contain a rodent.

If the relay clicks but the trap doesn't drop, don't worry. Your magnets are probably too strong for the electromagnets to counteract. Cut some small squares of tape and stack them on top of your permanent magnets until they're weak enough to be dislodged.

Your trap is done!

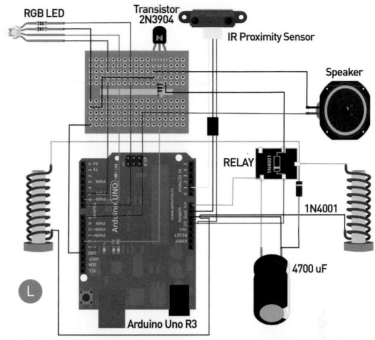

SMART RAT TRAP: WIRING DIAGRAM

RGB LED — Transistor 2N3904 — IR Proximity Sensor — Speaker — RELAY — 1N4001 — 4700 uF — Arduino Uno R3

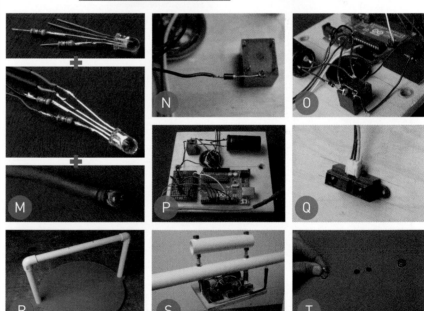

GET TRAPPING
Though designed for rats, the trap can be used to catch mice, squirrels, or other small animals. Place the trap in a rodent-infested area, sprinkle some attractive bait below the sensor (we've had success with combinations of chocolate, peanut butter, raisins, and nuts), and power it on. Happy trapping! ◉

Get more build photos and tips and share your build at makezine.com/projects/smart-rat-trap.

Written by Widar Hellwig ■ Illustrated by Megan Hellwig

"Light up" these targets and knock them down with IR LED guns

Infrared Shooting

ARCADE

WIDAR HELLWIG is an electronics enthusiast and electrical engineer who founded senselessdevices. com to share fun open-source projects.

MEGAN HELLWIG is a writer and illustrator who loves to create mad experiments. She lives in Daegu, South Korea.

LEARNING ABOUT TECHNOLOGY IS A LOT EASIER WHEN YOU'RE HAVING FUN — AND SHOOTING AT TARGETS IS FUN! With this amusement park–style arcade, you'll fire beams of infrared light instead of projectiles to trigger automated knock-down targets you can customize with soda cans, ducks, robots, or anything else you feel like toppling with a well-placed shot.

Here's how to build it from scratch. Or you can use my kits (senselessdevices.com/shop) and skip the programming steps entirely.

HOW IT WORKS
The toy gun emits invisible infrared (IR) light from an IR LED. The LED's light spreads at a wide angle, which makes aiming too easy — so we mount it deep inside a tube to collimate its light into a narrower beam.

This infrared light is similar to that of a light bulb or sunshine. To make the gun's light unique, we program a microcontroller to switch the LED on and off 38 thousand times a second (38kHz), a frequency that doesn't normally occur in background light.

Each target has an IR receiver that uses a phototransistor to see infrared light. When it detects the 38kHz signal from the gun, it triggers a servomotor to knock the target down.

1. SOLDER THE IR GUN'S ELECTRONICS
Cut a perf board 1¾"×½" so you've got 17 holes × 4 holes available. Install the components on one side — the speaker (for sound effects), infrared LED, red LED (for a gun flash effect), 0.1µF capacitor, 27Ω resistors, and IC socket — then solder the connections on the bottom (Figure **1a**), following the schematic on the project page at makezine.com/infrared-shooting-arcade.

Connect the 2xAAA battery holder: black lead to GND (pin 14 on the ATtiny24) and red to pin 1.

Solder two 6" wires to the trigger pushbutton, then solder these to the board too (Figure **1b**).

2. BUILD A PROGRAMMING ADAPTER
To program your own chips, you'll need an adapter to connect the AVR ISP programmer to the ATtiny24 microcontroller. You can get an adapter kit from Inside Gadgets (insidegadgets.com) or build one from scratch following the instructions on the project page online.

3. PROGRAM THE GUN MICROCONTROLLER
Connect the AVR ISP programmer to your adapter

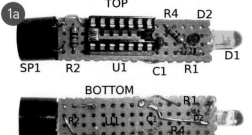

1a

TOP
R4 D2
D1
SP1 R2 U1 C1 R1

BOTTOM
R1
R2 U1 C1 D2
R4

Widar Hellwig

and insert an ATtiny24 chip into the adapter, with its notch oriented closest to the socket's lever.

Download the project code (see "Tools" list) and open the file *irgunREV100.hex* in your AVR programming software. Plug the programmer into your computer, then upload the code to the chip. Finally, insert the ATtiny24 into the IC socket on the gun circuit board.

1b
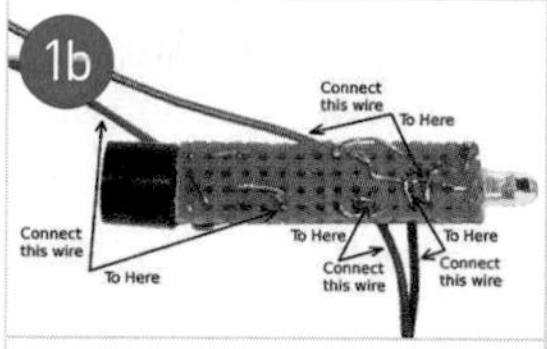

3
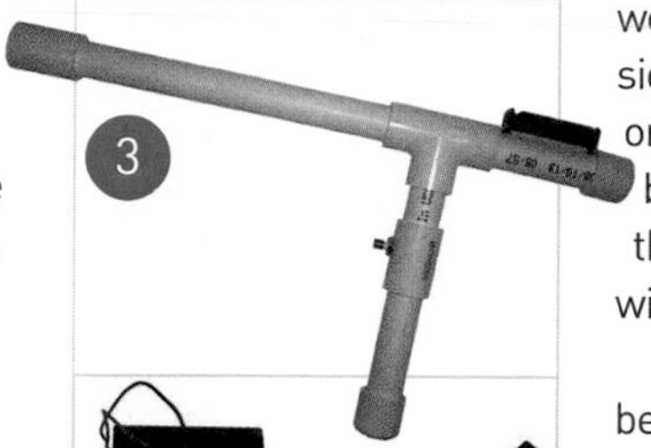

4
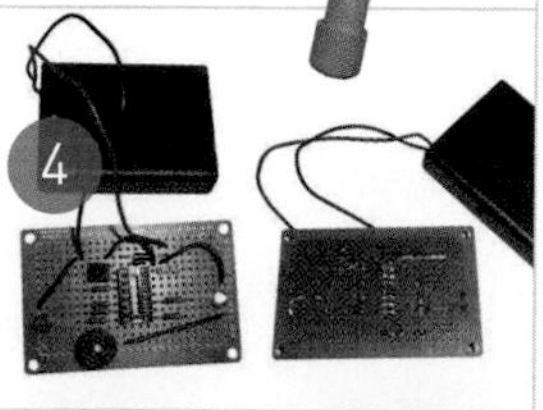

6a

6b

7

4. BUILD THE GUN

You can mount the electronics in an existing toy — an old Nintendo gun already has a trigger switch — or make a DIY gun from ½" PVC pipe and fittings as shown here.

Mount the electronics inside the tee fitting, then drill a 5⁄16" hole in a coupler to install the trigger pushbutton. Notch the back tube to accept the battery wires, then press on the end cap.

Verify that the IR LED is pointing straight out the barrel. Stick the battery holder onto the back with double-stick tape, and press-fit the gun together (Figure 3).

5. SOLDER THE TARGET ELECTRONICS

The target uses another ATtiny24 to operate a speaker, two LEDs, and a servomotor. When the IR receiver detects a shot from the IR gun, the microcontroller produces a "ding" sound effect with the speaker, flashes the white and red LEDs rapidly, and sends pulses to move the servo 90° and then return it. Use the servo to move a flag down and up again or to knock a can down.

Following the schematic online, solder the IC socket to the proto board and install the IR receiver module, followed by the 100Ω resistors and then the speaker, capacitor, and red and white LEDs. Install the diode and servo connector, and solder in the ground jumper wires, then connect the 4xAA battery holder (Figure 4).

6. PROGRAM THE TARGET MICROCONTROLLER

Program the ATtiny24 microcontroller using the other hex file, *irtargetREV100.hex*, and install it in the IC socket. Then insert batteries into the battery holder.

7. BUILD THE MECHANICAL TARGET

For a simple target, cut a small wood block and use double-sided tape to stick the servo to one corner and stick the circuit board to the front edge. Secure the battery pack and loose wires with a rubber band.

For a more rugged and better-looking target, use a project box and drill clearance holes for the LEDs, IR receiver, and speaker (Figure 6a).

Print an image to use for a target and tape it to a drinking straw. Attach the straw to the horn of the servo. Your target is ready; power it up and test it by shooting at the paper target. The IR beam should be wide enough to trigger the target's sensor when you aim at the paper target.

To knock down another object such as a tin can, mount the servo horizontally and attach a stick to strike the can (Figure 6b).

8. CREATE YOUR SHOOTING GALLERY

Place the targets on shelves and decorate your gallery to look like an amusement park arcade. You can do this quickly with cardboard and drawings, or even sew custom cloth curtains with embroidered signs (Figure 7).

My infrared shooting arcade first appeared at Maker Faire Bay Area 2014. You can use it to liven up all kinds of events, from birthday parties to fundraisers to Saturday afternoons. Don't be afraid to get creative: Place targets throughout a room or house; or hide them on furniture or shelves for kids to find and shoot at. Just move the targets occasionally to completely refresh the game, or try longer gun barrels to make aiming more challenging.

For more tips, tutorials, part numbers, and schematics, visit makezine.com/infrared-shooting-arcade.

Time Required: A Weekend
Cost: $25–$150

Materials

TO BUILD ONE GUN AND ONE TARGET:

- **PC boards, 371-hole**
- **Perf board** RadioShack #2761396
- **Microcontroller IC chips, Atmel ATtiny24, DIP package (2)**
- **IC sockets, 14-pin DIP (2)**
- **Resistors: 27Ω (3), 100Ω (6)**
- **Infrared LED** RadioShack #2760142
- **Infrared receiver** Mouser Electronics #782-TSOP32138
- **Servomotor, micro size**
- **Speakers, 12mm (2)**
- **LEDs, high brightness: red (2) and white (1)**
- **Switch, momentary pushbutton**
- **Capacitors, 0.1µF (2)**
- **Header, 3-pin**
- **Diode, 1N4004**
- **Battery holders: 4xAA (1) and 2xAAA (1)**
- **Hookup wire, 22 AWG**
- **Project box (optional)**
- **PVC pipe, ½", 2' length**
- **PVC fittings, ½": tee (1), couplers (2), end caps (2)**
- **Foam mounting tape, double-sided**
- **Old bookcase or shelves**

TO BUILD AN ATTINY24 DIP ADAPTER:

- **Power supply, 9V 300mA**
- **Voltage regulator, 5V**
- **Capacitors, 10µF (2)**
- **Resistor, 10kΩ**
- **DIP socket, 14-pin**
- **Header, 6-pin**
- **Prototype board**

Tools

- **Soldering iron and solder**
- **Wire cutters**
- **Pliers, needlenose**
- **Handsaw or PVC pipe cutter**
- **Drill and bits**

OPTIONAL, IF YOU'RE PROGRAMMING YOUR OWN CHIPS:

- **AVR ISP programmer**
- **Computer with AVR programming software** such as AVRStudio (atmel.com) or AVRDUDE (free from nongnu.org/avrdude)
- **Project code** Download *irgunREV100.hex* and *irtargetREV100.hex*, free from senselessdevices.com/arcade.html.

Build-a-Bin
Customizable Picking Bins

Get organized! Instantly design custom-sized bins, then cut them by laser or by hand.

Written by Dan Royer

DAN ROYER, with your help, is going to put construction robots on the moon.

Find out more on his website marginallyclever.com or come say hi in person at World Maker Faire in New York.

Time Required:
30–60 Minutes
Cost:
$0–$20

Materials

» **Cardboard, bristleboard, heavy cardstock, or corrugated plastic** aka Coroplast
» **Glue or tape (optional)**

Tools

» **Computer** with internet connection
» **Laser cutter (optional)**
» **Scissors**
» **Ruler and hobby knife** if you don't have a laser cutter

I FUSS A LOT ABOUT HOW TO ORGANIZE TOOLS AND PARTS SO THAT PEOPLE CAN EASILY FIND AND USE THEM. This is a constant challenge in my day-to-day business selling robot kits, as much as it is at my local hackspace. What I've figured out so far is that everything has to have a dedicated home address that's very visible and easy to access. "Picking bins" hit a sweet spot that meets a lot of needs.

They're stackable; they're open in the front; they've got room for a label on the front; there's no lid to lose; it's hard to put them on a shelf the wrong way; they can be made without glue, tape, or fancy tools; they can be made from recycled materials; and you can see what's inside without opening anything.

Inspired by Uline picking bins and Rahulbotics' online box generator, I created Build-a-Bin, a web page that generates the plans for you to make similar bins in your favorite sizes.

1. DESIGN YOUR BINS

Here's a sketch of a basic bin (Figure A). For your bins, you need to decide the bin height, width, and depth. Build-a-Bin automatically calculates the height and depth of the opening.

If your workspace is like mine — small and/or cluttered — you might want to design a variety of bin sizes that can stack on each other.

2. INPUT YOUR BIN DIMENSIONS

Go to the Build-a-Bin web page at marginallyclever.com/other/build-a-bin.php and input your bin's dimensions, along with the thickness of the material you'll be cutting (Figure B). Build-a-Bin will automatically draw a diagram of your bin to match your dimensions. Black lines show where to cut, red lines show where to fold.

Write down your dimensions, as you might want to make more of the same size later.

> **NOTE:** Build-a-Bin does not check for sane input values. It is possible to create nonsense designs. Use at your own risk.

3. SAVE THE DXF FILE

Build-a-Bin also generates a drawing file in the DXF format, which can be read by most laser cutter software. (You can also print the DXF drawing to make templates for hand cutting.)

You'll need to save this DXF file to your computer. Just copy the text from

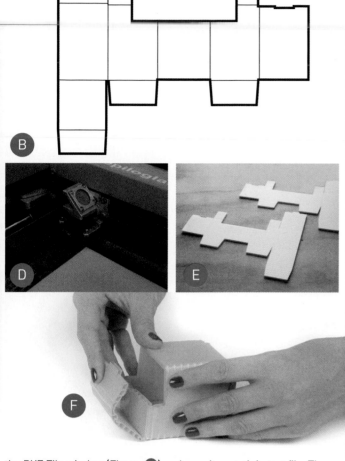

the DXF File window (Figure **C**) and save it as a plain text file. Then change the file extension from *.txt* to *.dxf* and you're good to go.

4. CUT YOUR BINS

Send the DXF file to your laser cutter and cut out as many bins as you like (Figures **D** and **E**).

If you don't have a laser cutter, open the DXF file in Inkscape and print it out, then use it as a template to cut your bins by hand.

5. SUPER MAGICAL ORIGAMI TIME!

Cut along the black lines, crease and fold on the red lines. Slot the back of the box together first, then the top, and finally the lip on the front (Figure **F**), fitting its tab through the matching hole. Voilá!

Label the front and stack on your shelf. I like to label mine like Battleship squares (A3, B5, etc.) so if they get moved it's obvious and they'll find their way home.

6. NOW MAKE MORE

Corrugated cardboard is ideal for these bins; cheap and recyclable. You can also use bristleboard or heavy paper cardstock. The *Make:* Labs had good results cutting Coroplast corrugated plastic. It's a little harder to fold but it's tough and durable and it looks great. ◑

+ Special thanks to Evan Jones and Brian Melani for helping to improve the Build-a-Bin script.

See more photos and show us your bins at makezine.com/build-a-bin. Tweet a picture of your bin and tag it #buildabin.

3 Fun Things to CNC

We asked our friends at Other Machine Co. (othermachine.co) for fun projects they like to cut with their new Othermill, a desktop CNC mill for circuit boards and other small projects.

1. CHOCOLATE MILLENNIUM FALCON

Ed Lewis

instructables.com/id/Chocolate-Millennium-Falcon

Chocolate may not be the first thing that comes to mind when you think CNC, but with CAD files, machining wax, and food-grade silicone, you can mill a Millennium Falcon chocolate mold that'll make Han Solo swoon. *Pew pew nom nom!*

2. PCB HUMMINGBIRD NECKLACE

Sam DeRose

instructables.com/id/PCB-Hummingbird-Necklace-on-the-Othermill

Cut this sweet little bird straight out of a blank PCB and etch the circuit on one side and a decorative pattern of your choosing on the other. Bonus LED backlighting brings ambiance on the go.

3. INTERACTIVE DIGITAL DISCO BALL

Colin Willson

instructables.com/id/Digital-Disco-Ball

Mill out custom-shaped circuit boards that actually fit together to form a futuristic disco ball. Inside, distance sensors connected to an Arduino change the brightness of specific LEDs according to how much activity they detect. No parking on the dance floor.

Backyard Climbing Wall

Build a real training wall that doubles as an awesome kids' playground

BENTON CALHOUN is a father of four who loves sharing outdoor activities with his kids. In between, he's a professor of electrical engineering and an entrepreneur.

Time Required: A Couple Weekends

Cost: $400–$450

Materials

- » Dimensional lumber, posts: 6×6 and 4×4
- » Dimensional lumber, framing: 2×4, 2×8, and 2×6
- » Joist hangers, 2×6
- » Clay and/or concrete
- » Carriage bolts
- » Plywood sheets, ¾"
- » Primer
- » Exterior paint
- » Sand
- » T-nuts and bolts, ⅜"
- » Climbing holds and anchors
- » Plywood, ¼"
- » Sheet tin
- » Steel pipe, ¾"

Tools

- » Circular saw
- » Drill and drill bits
- » Shovel, post-hole digger, or auger
- » Paint rollers or brushes

Written and photographed by Benton Calhoun

I CLIMBED IN HIGH SCHOOL AND COLLEGE, BUT HAD TO CUT WAY BACK ONCE MY WIFE AND I HAD CHILDREN. It was frustrating to be so out of shape. For conditioning my forearms, for training climbing moves, and just for fun, I've wanted a climbing wall for a long time. But it's tricky to find space for an indoor climbing wall — and to convince my wife that it's not going to seriously detract from the decor. So I finally decided to try building an outdoor wall. I gave myself the additional challenge of making it double as a kids' play structure.

I thoroughly enjoyed building it — mostly in 1- to 3-hour chunks at night after the kids went to bed, using floodlights from the house and a headlamp. If I'd done it all at once, I think I could've finished in 3 or 4 days. The kids and I have gotten years of fun out of it, and it's still going strong.

1. PLAN IT

I made a rough SketchUp drawing to play with different angles for my climbing panels (Figure Ⓐ) so they'd provide a variety of "terrain," and thought about how to incorporate a two-tier kids' fort on the back, with monkey bars and a slide.

Then I made a detailed SketchUp model to plan the major supports (Figure Ⓑ) and help me write a bill of materials. You can grab it at makezine.com/go/climbing-wall and play with it in 3D.

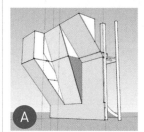
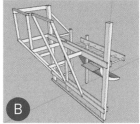

2. SET POSTS AND MAJOR SUPPORTS

Dig holes 4' deep for the 6×6 posts, and 2'–3' deep for the 4×4 posts.

It's tricky to set the big 6×6s in the hole; so you may want to ask a friend to help. I used 2×4s nailed to stakes to align the posts vertically, then tamped clay soil back into the hole to firm up the post. I also used some concrete for the big posts.

Attach the 2x8 cross braces.

3. FRAME IT

For the overhangs, use metal hangers to attach 2×6 "joists" to the 2×8s. First cut these to length and then cut angles to match the angled 2×4 wall studs. The trickiest part is cutting the notched bottom of the angled 2×4s so they'll sit on the lower 2×8s.

Once everything is screwed together, strengthen the key joints with carriage bolts.

4. PREPARE THE PANELS

Use a spade bit to drill holes for ⅜" T-nuts where you will place your climbing holds. I laid out a grid of 8" squares, then offset the holes 0" to 3" to get some pseudo-random patterns while keeping good hold spacing.

For weatherproofing, paint the panels on both sides with primer and textured paint. Just a spoonful of sand in a gallon of standard exterior paint gave the right feel for me.

Finally, pound in the T-nuts from the back. I recommend stainless steel if you can afford it. I used zinc T-nuts, and after 5 years, rust has made some of my holds impossible to move.

5. MOUNT THE WALLS AND HOLDS

Screw the panels to the framing. I started at the bottom and worked up.

Get some holds. Look online; I bought most of mine on eBay. Bolt the holds on the wall and get climbing. I found that more holds = more fun.

I also bolted a chain anchor at the top of each panel (through the 2×8) to support top roping, although we mainly boulder.

6. FINISH THE PLAY STRUCTURE

The main fort platform has a ladder and a back entrance behind the big overhang. A smaller ladder leads to the top platform, which is protected by a section of premade deck railing.

I made monkey bars from ¾" metal pipe set between two 2×4s with blind holes drilled halfway into them. And I made a slide by bending 3 pieces of ¼" plywood then covering them with a sheet of tin. It's loud, but it's fast it and works great! ◗

For more construction tips and photos, visit the project page at makezine.com/backyard-climbing-wall.

Workshop Light Doorbell

Can't hear the door? Hack a wireless doorbell to turn on a light too. *Written and photographed by Jason Poel Smith*

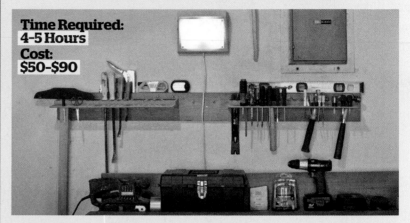

Time Required:
4–5 Hours
Cost:
$50–$90

Materials

» **Plug-in wireless doorbell, AC powered**
» **Extension cord**
» **Light socket**
» **Light bulb, low-wattage CFL**
» **Wire nuts, twist-on**
» **Relay, 12V solid state**
» **Quick-disconnect connectors**
» **Switches, sliding (2)**
» **555 timer IC**
» **Resistors: 1kΩ (1), 100kΩ (1)**
» **Capacitors: 330µF (1), 100µF (1)**
» **Printed circuit board**
» **Jumper wires**
» **Heat-shrink tubing**
» **Project enclosure**

Tools

» **Screwdriver**
» **Soldering iron and solder**
» **Knife**
» **Wire cutters/ strippers**
» **Hot glue gun (optional)**
» **Rotary cutting tool (optional)**

WHEN I'M USING POWER TOOLS IN MY WORKSHOP, I can't hear the doorbell ring. So I modified a wireless doorbell system to turn on a light in addition to playing a tone at my workbench. Here's how to do it.

1. HACK THE WIRELESS DOORBELL

Extend the speaker wires, connect a lead to negative DC power, and solder the 2 switches to the output transistor.

2. BUILD A CONTROL CIRCUIT

My circuit uses a 555 timer chip in monostable mode, powered by the doorbell unit. When pin 2 is triggered by the doorbell, pin 3 will switch on a relay for a certain amount of time, based on resistor and capacitor values you choose.

3. CONNECT THE RELAY AND LAMP

Solid-state relays require little power to operate. Connect the 555's pin 1 to the relay's negative input, and pin 3 to the relay's positive input. Slice the extension cord to connect the relay and the light socket.

4. MOUNT IT ALL

I put a translucent front panel on an old project box. Plug the doorbell into the extension cord, and then into the wall.

Now when someone rings your doorbell, the light turns on. Use the switches to toggle the tone or light on or off. Silent mode (light only) is good when someone's asleep. ◗

For more construction tips and photos, visit the project page at makezine.com/light-doorbell.

JASON POEL SMITH has an undergraduate degree in mechanical and electrical engineering and spends most of his time chasing his new baby and making the video series "DIY Hacks and How-Tos." youtube.com/make

Tracking Planes with RTL-SDR

Written by
David Scheltema

Follow commercial flights and map their exact locations in the sky with an inexpensive software-defined radio

Time Required:
1 Hour
Cost:
$65–$100

Materials

» **BeagleBone Black single-board computer** Maker Shed item #MSGSBBK2, makershed.com

» **RTL-SDR software-defined radio** with RTL2832U chipset, such as Adafruit #1497

» **Power supply, 5V 1A, or USB mini cable**

SOFTWARE-DEFINED RADIOS (SDR) ARE GAINING IN POPULARITY, AND IT'S NOT HARD TO SEE WHY — using them, your computer can tune into in an enormous range of frequencies, including FM radio, unencrypted police and fire bands, aircraft transponders, and in many countries, digital TV. The most popular SDR devices for the money are known as RTL-SDR because they're based on the Realtek RTL2832U, a demodulator chip that supports the USB 2.0 interface.

With an inexpensive RTL-SDR USB dongle and properly configured software, you can track commercial airplane flights and output their locations to mapping software to see exactly where they are in the sky. In this project you'll

learn how to do just that, using a very affordable single-board computer, the BeagleBone Black. While this build is not an original or exhaustive account of the technology, it's a useful aggregation and an example of the amazing things software and specific hardware can accomplish.

There are two main software packages to configure. First, the drivers for the RTL-SDR USB dongle, which require very little configuration, just installation. And second, dump1090, a program that tunes your SDR to 1090MHz, collects the data and outputs it on a locally hosted website.

1. PHYSICAL BUILD

The build is very straightforward. Attach the antenna to the SDR dongle, then plug the dongle into the BeagleBone Black's USB port.

Connect an Ethernet cable to the RJ-45 jack and a 5V power supply to the barrel jack. If a barrel jack supply is not handy, power can be supplied over the micro-USB port, but remember, amperage is limited at 500mA per USB specification.

That's it! The physical build is done.

2. SSH AND UPDATE DEBIAN

The BeagleBone Black should be booted and the blue LEDs flashing quickly. Rather than connecting a monitor and keyboard to the device, connecting remotely over SSH is preferable.

WINDOWS USERS:

Download and install an SSH program such as Putty. Set up a connection to beaglebone.local as the root user. By default, there is not a root password set.

OSX AND LINUX USERS:

You already have an SSH client installed on your system by default, but you need to open a terminal session and type:

```
ssh root@beaglebone.local
```

If beaglebone.local doesn't work, use 192.168.7.2 instead. If a password prompt appears, simply hit return.

Now, regardless of platform you should have a command prompt on the Beagle. Refer to the first image if unsure.

Update the Debian package list:

```
apt-get update
```

and then (optionally) upgrade the system:

```
apt-get upgrade
```

3. LIST USB DEVICES

List any USB devices attached to the BeagleBone:

```
lsusb
```

The output should show an RTL device, such as:

```
Bus 001 Device 002: ID 0bda:2838 Realtek
Semiconductor Corp. RTL2838 DVB-T
```

If no device is listed other than the two Linux Foundation root hubs, make sure the RTL-SDR USB dongle is firmly inserted into the Beagle and try the command again.

4. INSTALL CMAKE AND LIBUSB

Debian doesn't come with a compiled package for rtl-sdr, so you'll need to build it from source. It's not difficult to do and the following steps walk you through all the command-line jargon.

First, install cmake, an alternative build system

that's used by a number of open-source projects.

```
apt-get install cmake
```

Building rtl-sdr requires a specific USB library. Install it like so:

```
apt-get install libusb-1.0-0-dev
```

5. CLONE THE RTL-SDR REPO

Clone the code from the git repository ("repo") of the project.

```
git clone git://git.osmocom.org/rtl-sdr.git
```

6. CONFIGURE AND BUILD

Change directories to the cloned repo:

```
cd rtl-sdr
```

Configure the build with cmake:

```
cmake ./ -DINSTALL_UDEV_RULES=ON
```

Then build with make, and install:

```
make
make install
```

7. CLONE DUMP1090

Change working directories to the root:

```
cd
```

Clone the dump1090 repository:

```
git clone https://github.com/MalcolmRobb/
dump1090 dump1090
```

Change directories to the *dump1090* cloned directory and compile:

```
cd dump1090/
make
```

8. NOW TRACK PLANES!

Run dump1090 in interactive mode, with the web interface running on port 8081:

```
./dump1090 --interactive --net --net-http-
port 8081
```

Now point a web browser to beaglebone. local:8081 to see the dump1090 web interface (Figure 8a).

You'll see a marker for each plane in the air, accurately plotted on a Google map! Clicking on the bogey shows flight information such as longitude, latitude, airspeed, and also the flight path the plane has taken (Figure 8b).

You can switch to a dark map scheme (Figure 8c) instead of the usual Google Maps colors or even use the alternative solution, OpenStreetMap.

I hope this project captures your imagination and demonstrates how powerful SDR-based projects can be. Watch for more cool SDR projects coming soon in *Make:*! ●

See complete step-by-step screenshots and share your RTL-SDR projects at makezine.com/projects/tracking-planes

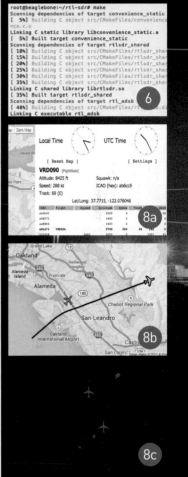

NOTE: By default dump1090 runs on port 8080; however, on the BeagleBone Black, the Apache webserver already runs on that port. That's why you told dump1090 to switch over to port 8081.

If for some reason beaglebone.local:8081 does not resolve, use 192.168.7.2:8081 instead.

Hep Svadja

Joseph Gay-Lussac and the
Technology of Fireproofing

Written by William Gurstelle ■ Illustrated by Peter Strain

USe SAFe, COMMON CHEMICALS TO MAKE DIY FIRE RETARDANTS

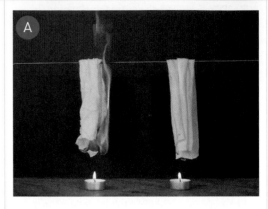

BEFORE HOMES WERE ELECTRIFIED, PEOPLE LIVED IN CONSTANT DANGER OF CLOTHING AND SURROUNDINGS CATCHING ON FIRE from the oil lamps (see this column in *Make:* Volume 22) and cooking fires used in daily life. As the study of chemistry became more rigorous, scientists began to think about how to protect people from this hazard. The person who first delved systematically into the problem was Joseph Louis Gay-Lussac, a 19th-century French polymath.

Gay-Lussac was a giant in the history of chemistry. Born during the turbulence of the French Revolution, the youthful Gay-Lussac moved to Paris to study with the eminent scientist Claude Berthollet. Berthollet was a close friend of the late Antoine Lavoisier, who had written the seminal book on chemistry. Gay-Lussac rose quickly to become a professor of physics at the Sorbonne, and while he is best remembered for investigating the properties of gases, his discoveries in inorganic chemistry make him the father of modern fireproofing as well.

In his Paris laboratory, Guy-Lussac documented a host of experiments that pioneered the art and science of making a combustible world a bit safer. The key to his success was his work with one particular substance: boron.

If sulfur, often called brimstone, is the element most closely associated with starting fires, then boron is the element most associated with stopping them. Boron is a good fire retardant because it chemically transforms the materials it treats, notably paper and fabric, inhibiting the spread of flame and promoting the formation of a protective layer of char that acts as a fire barrier.

Boron compounds have been widely used since antiquity, but pure boron itself was not isolated until 1808. At that time, Gay-Lussac and rival Humphry Davy, one of England's leading scientists (whom we met in this column in *Make:* Volume 20) were locked in a heated competition to isolate boron and thus lay claim to being its discoverer.

By 1808, Davy had discovered and named five elements — barium, calcium, strontium, sodium, and potassium — and felt confident he was close to isolating his sixth: the elusive boron. Word reached Gay-Lussac that Davy thought he was close. (In fact, many modern scholars believe Davy indeed isolated boron but was unable to prove it.)

So, across the English Channel, Gay-Lussac redoubled his efforts. Abandoning caution, he adopted a dangerous laboratory technique involving highly reactive pure potassium metal, and by taking this risk, soon isolated a substance he called "bore." Gay-Lussac was able to verify to the satisfaction of his peers that what he found was indeed a new element.

FIREPROOFING PAPER, CLOTH, AND WOOD

In 1821, when Guy-Lussac was experimenting with methods to make materials resistant to fire, he saturated fabrics with boron salts and found that boron compounds could indeed prevent cloth, paper, and other cellulose-based materials from burning (Figure **A**). His success in finding a fireproofing chemical that wouldn't affect the color of cloth — or turn it poisonous — was a breakthrough.

You can use Gay-Lussac's discoveries to make a combination fire-resistant hiking stick and campfire poker — a handy, multipurpose item that's ideal for overnight hikes into the wilderness. You'll use a slight variation of Guy-Lussac's original methods, but the principles are the same. As always, undertake this project at your own risk.

MAKE A FIREPROOF WALKING STICK/CAMPFIRE POKER

1. Go to a nearby forest and look for a downed branch that's 1"–1½" in diameter, relatively straight and without cracks or knots. A stick that reaches to your sternum is a good length.

Remove any small branches with a hatchet or saw. Remove the bark with a knife and file off any rough spots. If desired, you can drill a ¼" hole near the top end for a leather or rope loop.

2. Weigh and mix the boric acid powder and borax in a half-gallon of hot water in the bucket. Stir vigorously until the chemicals are completely dissolved (Figures **B** and **C**).

3. Place a piece of cotton cloth in the boric acid solution, saturating it thoroughly. Remove the cloth and hold it over the bucket, allowing the excess solution to drain back into the bucket. Hang the item up to dry (Figure **D**).

4. When the cloth is dry, test a corner by holding a match to it. The cloth should char and turn black but it should not actually ignite and burn (Figure **E**).

You can fine-tune the performance of your fireproofing solution by slightly adjusting the proportions of the borates and water.

5. When satisfied with your fireproofing solution, pour some in a tall, narrow container, and soak the end of your stick in it overnight (Figure **F**).

Remove the stick from the solution and let dry.

6. You now have an all-in-one hiking stick/campfire tending stick (Figure **G**). Use the stick to move firewood around in a campfire to improve flame height and fuel use.

Your stick is quite fire resistant (Figure **H**), but it's not entirely fireproof, so don't let the tip get too hot or it will eventually ignite. (But even then, I've found that this treatment will slow the spread of flames.)

How does boron prevent the wood or cloth from igniting? Gay-Lussac found that fire does not occur if air can be prevented from reaching the surface of organic materials by chemically coating the material fibers. When applied to wood, both boric acid and borax melt into a thin, glassy film that reduces the amount of volatile, flammable gas that the wood releases, causing the material to *calcine*, that is, thermally decompose into char — not actually go up in flames. **⊘**

What would you fireproof? Share ideas and tips on the project page: makezine.com/borate-fireproofing

Time Required:
1-2 Hours
Cost:
$5–$10

Materials

» **50g boric acid powder** from a hardware store. Boric acid is a weak acid that's deadly to roaches and ants, but safe for humans to handle.
» **60g borax** available at grocery stores
» **1gal hot water**
» **Cloth, 100% cotton** such as an old undershirt
» **Wooden stick, about 4' long**

Tools

» **Bucket, 3gal–5gal**
» **Spoon**
» **Long-handled lighter or fireplace matches**
» **Hatchet or saw**
» **Knife and file**
» **Drill (optional)**

WILLIAM GURSTELLE is a contributing editor of *Make:* magazine. His new book, *Defending Your Castle: Build Catapults, Crossbows, Moats and More* is now available.

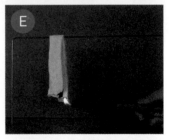

When Projects Fail:

My DIY wearable gadget never sold — but it led to my first article, a kit company, and a career in electronics and science.

Written and photographed by Forrest M. Mims III ■ Illustrated by James Burke

FORREST M. MIMS III
(forrestmims.org), an amateur scientist and Rolex Award winner, was named by *Discover* magazine as one of the "50 Best Brains in Science." His books have sold more than 7 million copies.

PROJECTS THAT FAIL ARE RARELY PUBLISHED — BUT THEY TEACH VALUABLE LESSONS THAT OFTEN LEAD TO SUCCESS. If you've spent much time designing and building projects, you know this well. I certainly do. Some of my failed projects made a major impact on my career in electronics and science.

During my senior year at Texas A&M University in 1966, Texas Instruments announced the development of a powerful LED that emitted several milliwatts of invisible near-infrared, about the same power output as a small flashlight. My great-grandfather had been totally blinded by a dynamite explosion when he was a young man, and the new LED gave me an idea for building a travel aid for the blind. So I hitchhiked to Dallas to meet Edward Bonin, one of TI's LED engineers.

The new LEDs cost $356 each, about $2,671 in today's money. Dr. Bonin said he would give me, a rank amateur, one of the LEDs if I could build a circuit that would generate the pulses needed to make the travel aid. I modified a 2-transistor Morse code practice oscillator board sold by a radio and TV repair shop for 99 cents and sent it to Bonin. He approved the circuit and sent it back, together with three of the sophisticated LEDs.

I quickly built and documented a prototype (**Figure Ⓐ**) and within a few days built a working travel aid that measured 2"×2"×4" (**Figure Ⓑ**). Flashes of invisible infrared emitted by the aid were reflected by objects up to 10 feet away. The reflected IR was detected by a silicon solar cell, and the resultant photocurrent was amplified by a transistor amplifier (salvaged from a hearing aid) and sent to an earphone, which emitted a tone. The closer the object, the louder the tone.

I tested the travel aid with more than 20 blind children and adults. It worked well, but the need to hold it in one hand was a drawback. Eventually I assembled the entire device on a pair of sunglasses. All the electronics were installed inside two ⅜"-diameter brass tubes mounted on the temples (**Figure Ⓒ**), the LED transmitter in one tube and the receiver in the other. A tiny hearing aid earphone in the receiver tube was coupled to the user's ear through a short length of plastic tubing.

The eyeglass aid worked well. It also received an Industrial Research 100 Award and a 1987 runner-up Rolex Award. But in the end, the project I had spent years developing was a failure.

The hearing aid companies I approached about manufacturing the travel aid responded that the potential liability was much too risky. What would happen if a blind user wearing the travel aid fell into a hole or was otherwise injured?

Though the travel aid was never manufactured, it taught me more about solid-state electronics and optics than my friends majoring in electrical engineering were learning. They were building old-fashioned vacuum tube circuits in their lab courses, while I was working with transistors and state-of-the-art infrared-emitting diodes.

The travel-aid circuits led to several new projects. I used the LED pulse generator circuit to flash a tracking light in night-launched model rockets I was flying to test a new kind of guidance mechanism (**Figure D**). After George Flynn, the editor of *Model Rocketry* magazine, watched one of those flights, he asked me to write an article about the light flasher. It was published in September 1969 (**Figure E**).

Ed Roberts and I were then assigned to the Laser Division of the Air Force Weapons Laboratory. We often talked about selling electronics kits through magazines like *Popular Electronics* and *Radio-Electronics*. When the light flasher article was published, we decided to form a company to build and sell light flashers and other model rocketry gear. We called it Micro Instrumentation and Telemetry Systems (MITS).

I eventually left MITS to pursue a new career as an electronics writer. Ed stayed and introduced a string of new products. I wrote the instruction manuals for some of them. I also introduced Ed to Leslie Solomon, the technical editor of *Popular Electronics*.

In 1974 Ed learned about the 8080, Intel's new 8-bit microprocessor, and soon began work on a microcomputer based on the new chip. The hobby computer era took off when Ed's Altair 8800 appeared on the cover of the January 1975 issue of *Popular Electronics*. When Paul Allen saw the magazine, he immediately bought a copy and took it to show his friend Bill Gates. They soon called Ed to say they were developing a version of BASIC for the Altair, and you know the rest of the story (see my column in *Make:* Volume 42, "The Kit That Launched the Tech Revolution").

Sometimes I wonder how this story might have ended had my great-grandfather not been blinded, or if TI hadn't invented the first infrared LED. Of course it's impossible to know in advance what might come from a failed or abandoned project — and that's motivation enough to press ahead.

GOING FURTHER

Have you developed a project that failed for technical or other reasons? Think about how it might have advanced your knowledge, and tell us at makezine.com/when-projects-fail.

And consider beginning a new project that's got only a marginal chance for success. My view is that every project is like a course in tech school or college, for the spinoffs from a failed project are sometimes as significant as those from the great successes. ◕

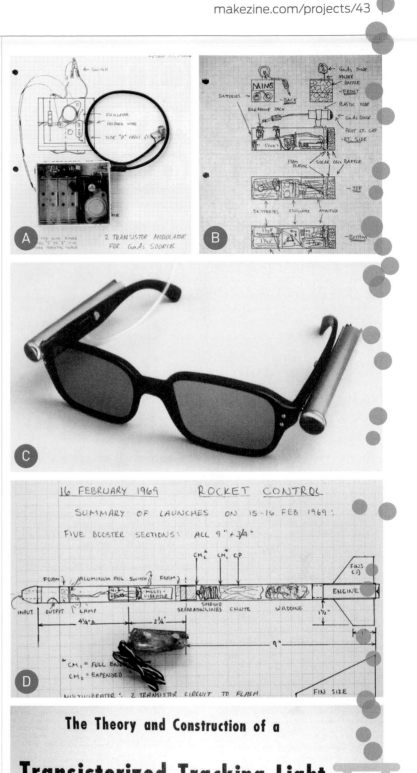

Pixilation: Full-Body Stop-Motion Animation

Time Required: 2-3 Hours Cost: $0-$50

Written by Michelle Hlubinka

MICHELLE HLUBINKA is director of custom programs for Maker Media, overseeing outreach and programs to help schools, kids, and families get into making.

You will need:

» **Digital camera** still or video
» **Computer with video editing software** Any will work. We like iStopMotion, free demo or $50 full version, boinx.com.
» **Tripod (optional)** but recommended
» **Paper and pen (optional)** for storyboards to plan your sequences
» **Props (optional)** such as chairs, costumes, cardboard boxes to hide in. Let your imagination roam.

TIP: To "fly," jump into the air when you take each picture. For a long flight, jump over and over while taking video, then delete unwanted frames.

TIP: Use colored tissue paper for flat planes of color, scrunched-up balls of color, or flattened versions of objects that get smooshed in your animation.

PIXILATION IS ONE OF MY FAVORITE KINDS OF ANIMATION, BOTH TO MAKE AND TO WATCH. Think of it as full-body, stop-motion animation. It'll get you up out of your comfy seats and active!

In stop-motion animation, like that of Ray Harryhausen, Wallace & Gromit, or Gumby, you animate an inanimate object by taking a picture, and then moving the object a little bit and taking another picture. You string these pictures or "frames" together in video-editing software to create the illusion of motion. People in pixilations appear to slide around without moving their legs, change places in the blink of an eye, or even fly. That's why it's called *pixilation* — it's about turning your subjects into magical pixies of a sort.

1. PLAN SOME SHOTS

Visit makezine.com/projects/pixilation for stop-motion videos to inspire you. Then incorporate these silly and surprising motions into your storyboards: Disappear/Reappear, Walk Through Walls, Magically Transforming Objects, Make Things Ooze or Blow Up, Sliding/Scooting, Flying.

2. DECIDE ON YOUR FRAME RATE

Experiment with 8 frames per second (FPS) for a faster shoot, or 24 FPS for smoother motion.

3. SET UP YOUR CAMERA AND STAGE

4. TAKE PICTURE #1

Put your actors in their places on the stage (this is called "blocking"), and take a picture.

5. MOVE, CLICK, REPEAT

Move everyone in the scene incrementally — changing their locations, moving a limb, etc. Take your second picture. Move again, take a third.

6. REPLACE THINGS

Magically swap one actor for another or for an object of the same size or color. Cover them with a coat or blanket, then shrink them to nothing!

7. WATCH YOUR VIDEO

Simply hit Play in iStopMotion. In other software, lay out your pictures on a timeline and specify their duration based on your frame rate.

8. SHARE

Upload your video and show the world! ◉

Get more tips and share your pixilation videos on the project page at makezine.com/pixilation.

PROJECTS

Keyboard
Refrigerator Magnets

Written by Jason Poel Smith ■ Illustrations by Julie West

SNAP OFF

CUT HERE

2

HOT GLUE

ROBOTS ROCK

I ALWAYS TRY TO FIND WAYS TO REUSE THE PARTS FROM MY OLD ELECTRONICS. I had a couple of old keyboards lying around, so I decided to use the keys to make alphabet refrigerator magnets.

1. REMOVE THE KEYBOARD KEYS

Use a narrow screwdriver to pry the keys off the keyboard. In most cases they will just pop right off. » Look at the backside of the keys. If the mounting tabs stick out past the body of the key, then you need to trim them so that you can mount the magnets flush with the backside. You can either cut the tabs with wire cutters or just break them off with needlenose pliers.

2. CUT THE MAGNETS INTO SQUARES

Now get some magnetic advertisements or magnetic business cards. Use scissors or a knife to cut the magnetic sheet into squares that are the same size as the back of the keys.

3. GLUE THE KEYS TO THE MAGNETS

Apply a large drop of hot glue to the back of one of the keys. » Then press the key onto one of the magnetic squares that you cut out. You want the bare magnet side to be facing out so that it can stick to the fridge better. » Repeat this process with all the keys. ◢

JASON POEL SMITH makes the "DIY Hacks and How Tos" project video series on *Make:*. He is a lifelong student of all forms of making and his projects range from electronics to crafts and everything in between.

You will need:

» **Computer keyboard**
» **Sheet magnets** new or recycled from magnetic advertisements, car signs, or business cards
» **Screwdriver**
» **Scissors**
» **Wire cutters**
» **Pliers, needlenose**
» **Hot glue gun**

Check out step-by-step photos at: makezine.com/projects/keyboard-refrigerator-magnets

Jason Poel Smith

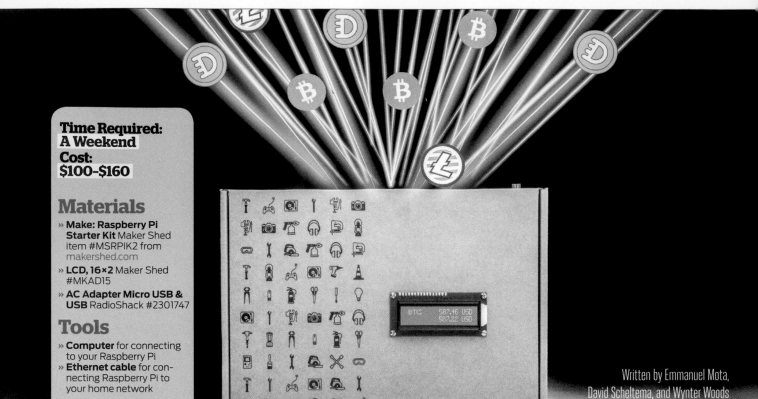

Time Required:
A Weekend
Cost:
$100–$160

Materials

» **Make: Raspberry Pi Starter Kit** Maker Shed item #MSRPIK2 from makershed.com

» **LCD, 16×2** Maker Shed #MKAD15

» **AC Adapter Micro USB & USB** RadioShack #2301747

Tools

» **Computer** for connecting to your Raspberry Pi

» **Ethernet cable** for connecting Raspberry Pi to your home network

Written by Emmanuel Mota, David Scheltema, and Wynter Woods

Crypto Currency Tracker
Know the current values of your favorite digital currencies in real-time

WHAT'S IN YOUR BITCOIN WALLET? IS DOGECOIN REALLY HEADED "TO THE MOON"? The Crypto Currency Tracker constantly monitors the value of three digital currencies — Bitcoin, Litecoin, and Dogecoin — and compares them against the U.S. dollar. You can build it easily using the inexpensive Raspberry Pi computer.

The tracker's software is built on Raspbian Linux and uses Python scripts to make JavaScript Object Notation (JSON) requests for each respective digital currency market index. The code pulls the currency values from the internet then displays both previous and current values.

1. BREADBOARD THE COMPONENTS
Insert the Pi Cobbler, 16×2 LCD, and a 10K potentiometer into a large breadboard. Then wire up the potentiometer and LCD following the detailed instructions at makezine.com/crypto-currency-tracker.

2. HOOK UP THE RASPBERRY PI
Connect the ribbon cable to the Raspberry Pi with its white stripe (pin 1) near the edge of the board. Insert an SD card flashed with Raspbian Linux. Finally, connect an Ethernet cable, HDMI cable, keyboard, and a USB micro cable with wall wart to power the Pi.

3. CONNECT TO THE RASPBERRY PI
Log in with username `pi` and password `raspberry`. Write down the IP address shown during boot-up. You can use it to install and configure software over SSH.

4. INSTALL THE PYTHON ENVIRONMENT
Connect to the Pi directly (with keyboard, monitor, and Ethernet) or use SSH. Log in to the Pi, and from the command line execute:

```
sudo apt-get install python-dev
sudo apt-get install python-setuptools
sudo easy_install -U distribute
```

5. INSTALL GIT, PIP, AND RPI.GPIO
Install git, for cloning software repositories:

```
sudo apt-get install git
```

Install pip, a package manager for Python:

```
sudo apt-get install python-pip
```

Then use pip to install rpi.gpio, a Python module for controlling the Raspberry Pi's input-output pins:

```
sudo pip install rpi.gpio
```

6. INSTALL THE TRACKING SOFTWARE
Now you're ready to clone the Crypto Currency Tracker repository:

```
git clone https://github.com/Make-Magazine/wp14-raspberry-pi-crypto-currency-tracker.git
```

Change directories to what was just cloned:

```
cd ~/wp14-raspberry-pi-crypto-currency-tracker
```

and set up the 16×2 LCD library:

```
git submodule init
git submodule update –recursive
```

Finally, run the tracking software as root:

```
sudo ./crypto_currency_monitor
```

You're tracking! The project code is open source and easy to modify for following other currencies. So next time you want to appraise your digital coin, don't reach for your phone — use your own custom tracker. ◎

Get step-by-step instructions and video at makezine.com/crypto-currency-tracker.

Hep Svadja

Box Fan Beef Jerky

Written by Paloma Fautley ■ Illustrations by Julie West

JERKY IS THE ON-THE-GO SNACK THAT HAS FUELED NATIVE AMERICANS, PIO-NEERS, AND ASTRONAUTS ALIKE. And it's easy to make at home. This method, popularized by Alton Brown, uses an ordinary box fan and air filters to dry the meat.

1. SLICE IT
Buy your desired cut of beef (sirloin is a good choice), and slice it into thin strips using a long, thin blade.

2. MARINATE IT
Choose a marinade and let the meat soak in the refrigerator for 6–8 hours. Make sure to include ingredients that help the meat stay moist and tender. Honey is a good choice.

3. DRY IT
Remove the meat from the marinade and pat dry with a paper towel. Place the strips on top of a large, clean air filter (make sure it contains no fiberglass). Continue stacking meat and filters until you run out of meat. Place one final air filter on top and put the stack of filters on a box fan. Use a bungee cord to secure the filters, then turn on the fan.

Let the fan run and enjoy that sweet, sweet meat smell for 8–12 hours. Once the meat is dehydrated, enjoy! ◗

Check out step-by-step photos at: makezine. com/projects/box-fan-beef-jerky

PALOMA FAUTLEY is currently pursuing a degree in robotics engineering and has a wide range of interests, from baking to pyrotechnics.

You will need:
- » **Beef**
- » **Marinade**
- » **Knife**
- » **Kitchen container or plastic bags** for marinating
- » **Air filters, nonfiberglass (3–4)** for furnace or air conditioner
- » **Bungee cord**
- » **Box fan**

Paloma Fautley

Ultimaker

2

3

Eric Chu

Markus Seidt

ERIC CHU is a yo-yo hacker, robot builder, and industrial design student at California College of the Arts in San Francisco. He is a former *Make:* Labs engineering intern and 3D printer guru.

Sam Murphy

3D Printer Mods and Hacks

Better, stronger, faster — get the most from your printer

Written by Eric Chu

DESKTOP 3D PRINTERS KEEP GETTING BETTER as users keep coming up with clever upgrades. Try these three to get your machine printing its best.

1. HEATED GLASS BED

The right glass and where to get it

Whether you're printing on bare glass for PLA or with a layer of glue for ABS, parts seem to magically pop right off the glass when it has cooled — no prying or excessive force needed.

There are reports of successful printing on ordinary window glass, but it can crack due to thermal shock. Borosilicate glass, which is used for oven windows, is heat-resistant enough for most printable plastics. LulzBot, Airwolf 3D, McMaster-Carr, and a few other stores sell precut sheets for around $25 for a 200mm×200mm bed. Custom-cut sheets are more expensive and harder to find — there are only a few vendors and most require a price quote — but Voxel Factory (voxelfactory.com) in Montreal offers custom sheets for $44.

2. ASTROSYN STEPPER MOTOR DAMPERS

Less noise, more accuracy

These dampers from Astrosyn are essentially two steel plates held together by a hard rubber center. The rubber absorbs the noisy vibrations of the stepper motor and thus reduces the amount transferred to the chassis of the printer. The unit slips over a standard NEMA stepper (like the NEMA 17 shown here). Two threaded holes allow the damper to be screwed securely to your machine (I didn't have short enough screws, so I used some hex nuts as spacers).

I installed dampers to the x- and y-axis motors on an Ultimaker. There's a noticeable reduction in noise, though not as much as I'd hoped. Other printer designs may benefit more, particularly delta styles.

Astrosyn is based in the U.K., so it's harder to source these in the United States. I bought mine from a user in the "Delta robot 3D printers" Google Group for around $7 shipped each. There are also a few stores that sell these in the U.S. for a bit more.

3. CROSSFLOW FAN

New, effective way to cool prints

The crossflow approach to cooling prints is an experimental alternative method. Instead of mounting a little axial fan to the head of the printer, you mount a wide crossflow fan (commonly used in air conditioners) to the chassis of the printer and aim it over the entire print area.

Ultimaker forum members are testing this method with very positive results in PLA: better bridging, less warp on large parts, and sufficient cooling on small features. ABS seems to not benefit as much, and in some cases may suffer delaminations. Follow the thread in Ultimaker forums (makezine.com/go/crossflow) to learn more. ◗

Toy Inventor's Notebook

"OUIJA BE MINE" MAGIC MOVING VALENTINE CARD

Invented and drawn by Bob Knetzger

WANT TO SEND YOUR VALENTINE A REALLY SPECIAL MESSAGE THIS FEBRUARY 14TH? Here's an animated pop-up card you can make for Valentine's Day. When you slowly open the card, a hand sweeps across and moves the heart-shaped planchette over the Ouija board, spelling out the message: "HAPPY VALENTINE'S DAY!" It's both mysterious and romantic, perfect for any secret admirer to make and send.

It's also easy to make: Just go online to makezine.com/ouija-be-mine and download the 2 graphics files. Print them on cardstock, or laminate a paper printout onto thin cardboard. Cut out the hand and the long tab on the solid lines (don't forget to cut out the circular hole in the planchette). Use a small ⅛"-diameter punch to make the two holes in the hand and one hole in the tab. Score and slightly bend the tab on the dotted line.

Carefully score and fold the body of the card on the dotted line. Cut the slot and the small hole in the card with a sharp hobby knife. To assemble, use a brass brad to fasten the tab to the hand, then thread the tab through the slot in the card. Finally, fasten the hand to the card with a second brass brad. (The brads even look like 2 little brass buttons!)

Check the action by opening and closing the card and make any fine adjustments to the brads for smooth and easy movement. Well done, mystery Valentine! ◗

Download the graphics for printing, and see the finished card in action at makezine.com/ouija-be-mine.

EASY

Photography by *Make: Labs*

**Time Required:
3-4 Hours
Cost:
$10**

CHARLES PLATT
is the author of *Make: Electronics*, an introductory guide for all ages. He has completed a sequel, *Make: More Electronics* and is also the author of Volumes One and Two of the *Encyclopedia of Electronic Components*. Volume Three is in preparation.
makershed.com/platt

DIGITAL
POTENTIOMETERS
Create patterns in light and sound — no microcontroller necessary

Written by Charles Platt

Normally lurking unseen inside stereo systems, the digital potentiometer needs no adjustment, because it adjusts itself. Its fluctuating resistance can change the color or brightness of a light, the loudness or frequency of a sound, or any other parameters that depend on voltage and current.

STEPS ON A LADDER
Sealed inside the chip is a ladder of resistors. Figure Ⓐ shows the idea. The connections between the resistors are called "taps." If you have 127 resistors (as in this example), there are 128 possible taps, including those at the ends of the ladder. Digital potentiometers typically have 8, 16, 32, 64, 100, 128, 256, or 1,024 taps.

Two pins with the remarkably unimaginative names "A" and "B" provide access to the ends of the ladder, while a third pin, known as the Wiper, can be connected internally with any of the taps. Although the resistance between the Wiper and A or B changes in small, discrete steps, the transition is smooth enough for many purposes, such as adjusting the volume on a stereo.

Figure Ⓑ shows the pinouts of the AD5220 family of digital potentiometers (which include the AD5220BNZ10, AD5220BNZ50, and AD-5220BNZ100, having a total internal resistance of 10K, 50K, and 100K respectively). Pulses at

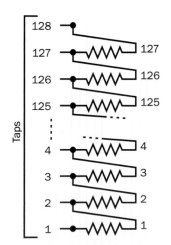

Ⓐ There is always one more tap than the number of resistance values in a digital potentiometer. In this case, 128 taps and 127 resistors.

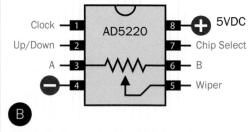

Ⓑ These pin functions are shared by all digital potentiometers in the AD5220 family.

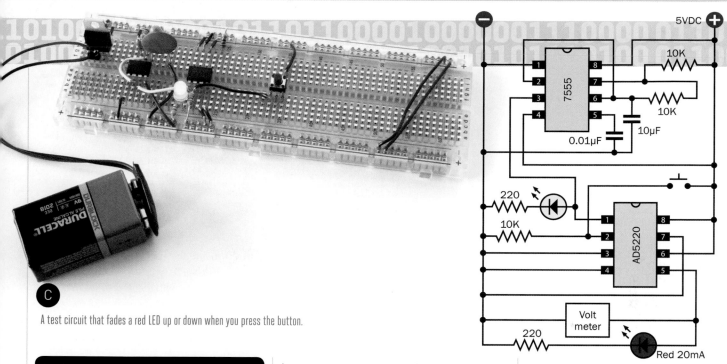

C

A test circuit that fades a red LED up or down when you press the button.

NOTE

You can also buy digital potentiometers in which each resistor has a numeric address, and a binary code tells the Wiper to jump directly to that address. However, this requires a serial communications protocol that is too complicated for me to deal with here.

the Clock pin move the Wiper connection one step at a time, while the logic state of the Up/Down pin determines whether the Wiper will advance toward A or B. The Chip Select pin must be grounded to activate the chip.

The AD5220 needs a 5VDC power supply, which you can provide using an LM7805 voltage regulator with a 9V battery. A and B are functionally identical, so you can apply voltage either way around, but the potential difference must never exceed 5V.

TIMING THE RESISTANCE

If you control the chip using pushbuttons, you'll have to debounce them to get rid of their voltage spikes. However, it's easier (and more interesting) to control the chip electronically, and I suggest using an Intersil 7555 for this purpose. A plain old 555 timer creates voltage spikes, which the digital pot may misinterpret as clock pulses.

Figure **C** shows a test circuit in which I connected the ends of the resistor ladder

(pins 3 and 6 of the AD5220) between the power supply and ground. As the Wiper moves between them, its voltage (on pin 5) will vary between 0V and 5VDC.

Begin with the red LED disconnected. Power up, and the yellow LED should flicker to confirm that the timer is sending pulses. The AD5220 always starts with the Wiper in the center of its range, and the Wiper voltage will gradually increase, because the Up/Down pin (pin 2) is connected through a 10K resistor to ground. You'll see your meter displaying the voltage as the Wiper climbs the ladder. Now hold down the pushbutton to connect pin 2 with 5VDC, and the output will count down almost to 0V (the actual value will depend on the internal resistance of your power supply).

Connect the red LED in parallel with your meter, press the button to make the chip count down, and the LED fades out completely around 1.6VDC. This happens because an LED, like any diode, requires a minimum voltage to function at all. I specified a red LED because it requires less forward voltage than other colors. A white LED, for instance, requires at least 3.2VDC. (These useful facts, and many more, are contained in my *Encyclopedia of Electronic Components*, Volume 2, which should be available around the time

Parts

» **Breadboard**

» **Jumper wires**

» **Power supply, 5VDC** such as an LM7805 voltage regulator with a 9V battery

» **Resistors, ¼W: 100Ω (3), 220Ω (6), 6.8kΩ (1), 10kΩ (10), 15kΩ (1), and 20kΩ (1)**

» **Capacitors, ceramic: 0.01µF (3), 1µF (3), 1.5µF (3), 2.2µF (3), and 10µF (3)**

» **LEDs, 20mA: red (1), green (1), blue (1), and yellow (3)**

» **Tactile switch**

» **Digital potentiometers (3)** Analog Devices AD5220BNZ10

» **555 timer ICs (3)** Intersil 7555

» **Counter ICs, dual 4-bit, 74HC4520 type (3)**

» **Transistors, bipolar, 2N2222 type (3)**

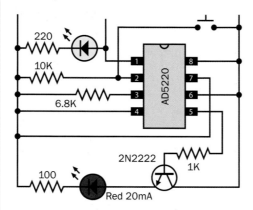

D The previous circuit has been modified to control a red LED through its full range of brightness.

you read this — see page 107 for more info.)

If you want the LED to remain dimly visible at the low end of the range, you need to add a resistor between pin 3 of the digital potentiometer and ground. And while you're at it, add a transistor, because we shouldn't really drive the AD5220 so close to its maximum rating of 20mA.

Figure D shows the lower half of the previous circuit rewired for this purpose. A 6.8K resistor now gives the Wiper a minimum of around 2VDC instead of 0V. This should stop the LED from going completely dark. Also, the series resistor for the red LED has been changed from 220 ohms to 100 ohms, because the transistor adds some effective resistance to the circuit. The current through my LED maxes out at 18mA and drops almost to 0mA. Use your meter to check if yours does the same. Now substitute a 1μF timing capacitor for the 10μF timing capacitor, and the LED should fade in and out quickly and smoothly.

NEXT STEP: FULL AUTOMATION!
Let's replace the pushbutton with something that will reverse the cycle automatically. The output from another (slower) timer could be applied to the Up/Down pin, but it would tend to drift out of sync with the first timer. What we need is a component that counts 128 cycles, then reverses the state of the Up/Down pin, counts another 128 cycles, reverses it again — and so on.

That sounds like a job for a counter chip! In fact an 8-bit binary counter runs through 256 cycles, which just happens to be 2 x 128. After the first 128 steps, the output changes from 01111111 to 10000000, so the most-significant-bit pin goes from low to high. It then stays high for another 128 cycles, at which point it changes from high to low. Just the thing! This is why I chose a digital potentiometer that has 128 taps, so we could make it step up and down with an 8-bit timer.

The final circuit and schematic are shown in Figure E, using a 74HC4520 timer chip, which contains two 4-bit counters, chained together. All of its outputs are unconnected except for the most-significant-bit pin.

MAKING IT MULTICOLORED
Build two additional copies of this circuit, one driving a green LED and the other driving a blue LED. Adjust the speed of the timers so that they are slightly out of phase. Now if you mix the light from the LEDs, it will run through all the colors of the spectrum in a pattern that seems random. You will have to increase the value for the 6.8K resistor for the green LED, so that the minimum voltage of the Wiper matches the minimum forward voltage of the LED. You'll need to do the same for the blue LED. This will be a matter of trial and error.

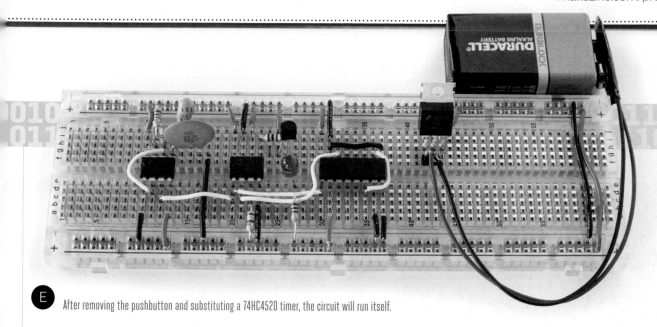

E After removing the pushbutton and substituting a 74HC4520 timer, the circuit will run itself.

NOW HEAR THIS

So far so good — but I'm only getting started. How about audio effects? Wire a 555 timer to run at an audio frequency. Instead of its timing resistor, insert a digital potentiometer (with a 5K additional series resistor, so that the total value never goes to zero). Add the 8-bit binary counter as before, and now you have a rising and falling musical tone.

The up-and-down cycle will quickly become boring, but my book *Make: More Electronics* tells you how to build a simple linear-feedback shift register (LFSR) to create a pseudo-random output of high and low states. Substitute this for the counter at the Up/Down pin of the digital potentiometer, and you'll have automated electronic music that sounds totally unpredictable. Alternatively, use the LFSR to add more randomness to your lighting circuit.

Other options are possible. The 10K version of the AD5220 can run at up to 650kHz. If you run it, conservatively, at 300kHz, it will cycle all the way up and all the way down almost 1,200 times per second. Since 1.2kHz is an audible frequency, you can connect the fluctuating output from the digital pot through an amplifier to a loudspeaker, and hear a very precise triangular sound wave. In the timer that sets the frequency, remove the timing resistor, substitute another digital potentiometer, control it with an LFSR, and you'll get a very different kind of random music.

Of course, digital potentiometers were never intended for this kind of weirdness. But that's what makes it so much fun. ⊘

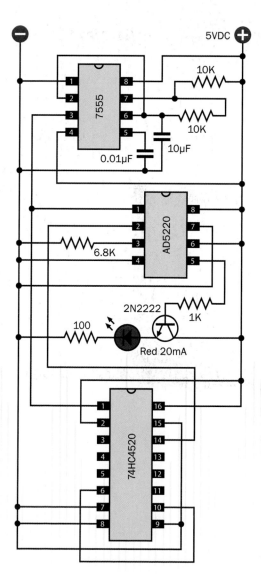

Mount your phone to a microscope for stunning close-ups

Written and photographed by Ben Krasnow

Extreme ZOOM

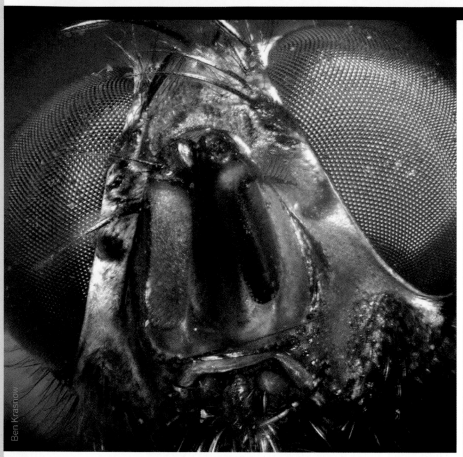

Ben Krasnow

I've always been fascinated by high-resolution photos of insects. The amount of detail at that small scale is impressive and also surprising because it surrounds us on a daily basis, but goes unnoticed without a bit of effort and a microscope. You don't need a fancy camera — even a cellphone camera will provide good results. With a little practice, you'll be able to capture and share your own high-quality images of the microscopic world.

Ideally you'll want a low-magnification, binocular microscope that provides light from above the specimen. But if you already have a conventional monocular microscope that provides light from beneath the specimen, you can easily use that as well.

CREATE ADEQUATE LIGHTING
Even if your scope already has a light that shines downward, you'll likely need to add more — the greater the light, the better the images. One solution is to take apart a desk lamp that has a circular fluorescent tube surrounding a magnifying glass. Remove the glass, and check that the light can be positioned so that it surrounds your specimen without interfering with the microscope itself (Figure Ⓐ).

Another option is to buy two or more goose-neck lamps, such as Jansjo from Ikea (shown in Figure Ⓓ). These lamps are really handy for lots of projects, and don't need to be modified for use with your microscope.

MOLD A CAMERA MOUNT
You can attach your cellphone or point-and-shoot camera to the microscope with a moldable plastic such as InstaMorph or Shapelock. These products come in pellet form and are softened by immersing them in hot water. The pellets form a blob that has the consistency of modeling clay, and can be wrapped around your microscope eyepiece and camera or phone. The plastic will retain its shape after it cools (Figure Ⓑ) and is not sticky, so the camera can be easily removed.

Be sure to have a properly lit specimen in

BEN KRASNOW works at Google[x], creating advanced prototypes, and previously developed virtual reality hardware at Valve. Follow Ben's personal projects on his YouTube channel, Applied Science: youtube.com/bkraz333

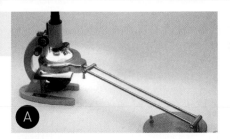

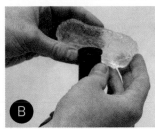

the microscope with the camera turned on while applying the moldable plastic. Very slight misalignments will cause some of the image to be lost. I made a small hook with the plastic for my cellphone mount, which holds the phone surprisingly well (Figures **C** and **D**).

If you have a camera with a removable lens (such as a DSLR), you'll get the best results by removing both the camera lens and the microscope eyepiece, and connecting the camera directly to the top of the microscope. This can be accomplished by disassembling a broken camera lens and salvaging its lens mount to help couple the microscope barrel to the camera with moldable plastic (Figure **E**), or by using a machined or 3D-printed adapter (there are lots of shared designs online at makezine.com/go/microscope-adapters).

ADJUST CAMERA SETTINGS

Your camera's focus should be locked to infinity if possible. This way, the knob on the microscope will control the focus, and the camera's autofocus mechanism will not continually try to adjust.

The camera's exposure should also be set manually so that multiple images have the same brightness.

Many cameras also allow a custom white balance to be set through a menu function, which improves overall quality. You can do this by placing a piece of white paper under your microscope, and activating the function with all of your lighting positioned as it normally would be for a real specimen.

Using a wireless or wired remote control will allow you to take photos and turn the focus knob on the microscope without disturbing the camera at all.

CAREFULLY COLLECT SPECIMENS

You can mount insects and other objects very effectively with a pin and a piece of modeling clay or putty (Figure **F**).

You can check windowsills for dead insects, but if you want to photograph a living insect that

you've found, it can be difficult to kill it without damaging it. One method is to place the insect in a small jar and carefully spray in some gas from a canned "duster." The gas will displace the oxygen in the jar, and kill the insect as humanely as possible without ruining its detail.

SHOOT MULTIPLE IMAGES

Once you're taking photos of tiny objects, you'll notice that very little of the image is in focus at any given time, like this tiny purple lanthanum hexaboride crystal (Figure **G**).

This is a physical limitation of the optics, but it can be overcome with a clever trick: Take 10 or more photos of the specimen at different focus settings, and combine them so that all of the sharp parts of each photo are combined into an image that is completely in focus.

This technique is called *focus stacking*, and can be accomplished with Adobe Photoshop or a free program such as CombineZM. Set up your camera as described above, then adjust the microscope's focus knob so that the highest part of the specimen is in focus. Take a picture, then turn the focus knob so that a slightly lower part of the specimen is in focus, and take another picture. Continue this until the bottom of the specimen is in focus. Copy all the images to your computer, and use the focus-stacking program to combine them into a sharp image (Figure **H**).

I hope these techniques allow you to explore the microscopic world and share your creative vision of it. You may find new ways to photograph insects, ice crystals, plants, rock formations, and other everyday wonders. ✪

NEED A MICROSCOPE?

If you don't have a microscope, you can purchase a suitable model for less than $100 new or from an auction of used equipment. The Amscope SE100-ZZ is an economical model that fits this description. When taking photos, you'll only use one eyepiece, but the binocular feature is nice when simply looking through the scope.

HOW TO USE A
BREADBOARD!

Written by Sean Michael Ragan ■ Illustrated by Jody Culkin ■ Circuits by Anne Mayoral

+ Get the full, 10-circuit version of the *How to Use a Breadboard* PDF and companion kit at makershed.com.

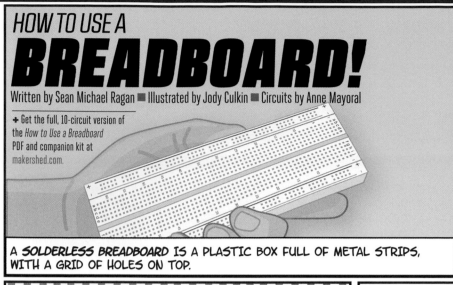

A **SOLDERLESS BREADBOARD** IS A PLASTIC BOX FULL OF METAL STRIPS, WITH A GRID OF HOLES ON TOP.

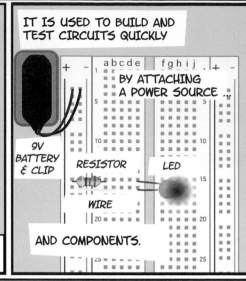

IT IS USED TO BUILD AND TEST CIRCUITS QUICKLY

BY ATTACHING A POWER SOURCE ...

9V BATTERY & CLIP   RESISTOR   LED

WIRE

AND COMPONENTS.

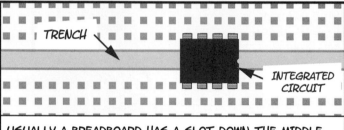

TRENCH

INTEGRATED CIRCUIT

USUALLY A BREADBOARD HAS A SLOT DOWN THE MIDDLE CALLED A **TRENCH**. THE WIDTH IS DESIGNED SO MANY **INTEGRATED CIRCUITS (ICS)** FIT RIGHT ACROSS IT.

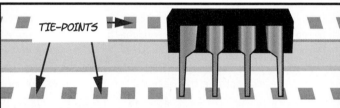

TIE-POINTS

THE HOLES, CALLED **TIE-POINTS**, ARE THE SAME DISTANCE APART AS THE PINS ON MANY **ICS** AND OTHER COMPONENTS.

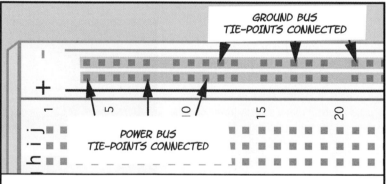

GROUND BUS TIE-POINTS CONNECTED

POWER BUS TIE-POINTS CONNECTED

A BREADBOARD'S LONG EDGE USUALLY HAS **TWO DISTRIBUTION BUSES** FOR CONNECTING **POWER** AND **GROUND**.

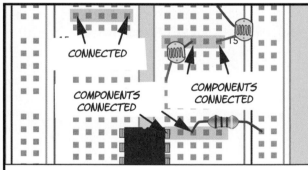

CONNECTED

COMPONENTS CONNECTED

COMPONENTS CONNECTED

ROWS OF CONNECTED HOLES RUN PERPENDICULAR TO THE BUSES. TO CONNECT COMPONENT LEADS, PUT THEM IN THE SAME ROW.

SHORT PIECES OF WIRE CALLED **JUMPERS** ARE USED TO CONNECT DIFFERENT ROWS.

POWER AND GROUND JUMPED TO BUSES ON OTHER SIDE OF BOARD

IC PIN CONNECTED TO GROUND

IC PINS CONNECTED TOGETHER

USING A SOLDERLESS BREADBOARD ALLOWS YOU TO GET YOUR CIRCUIT UP AND RUNNING QUICKLY SO YOU CAN TEST IT. ONCE YOU HAVE IT JUST RIGHT, YOU CAN BUILD A MORE PERMANENT VERSION ON PERFBOARD OR A PCB!

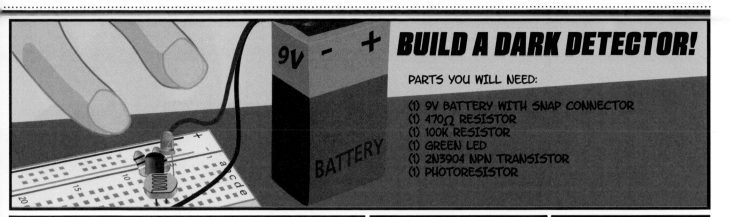

BUILD A DARK DETECTOR!

PARTS YOU WILL NEED:

(1) 9V BATTERY WITH SNAP CONNECTOR
(1) 470Ω RESISTOR
(1) 100K RESISTOR
(1) GREEN LED
(1) 2N3904 NPN TRANSISTOR
(1) PHOTORESISTOR

THE TRANSISTOR WILL TURN ON THE LED WHEN THE VOLTAGE TO ITS BASE LEAD GOES HIGH. IN THE LIGHT, THE PHOTORESISTOR HAS A LOW RESISTANCE, AND THE TRANSISTOR BASE VOLTAGE STAYS LOW.

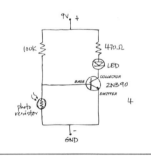

IN THE DARK, THE PHOTORESISTOR HAS A HIGH RESISTANCE, CAUSING THE TRANSISTOR BASE VOLTAGE TO GO HIGH, ALLOWING CURRENT TO FLOW THROUGH THE LED.

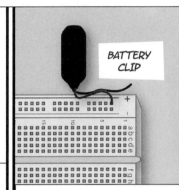

FIRST, CONNECT THE CURRENT SOURCE (IN THIS CASE, THE RED LEAD OF A 9V BATTERY CLIP) AT THE "TOP" OF THE CIRCUIT.

HOOK UP TWO CURRENT-LIMITING RESISTORS TO PROTECT THE COMPONENTS FROM DAMAGE: A BIG ONE (100K) FOR THE TRANSISTOR BASE AND A SMALL ONE (470Ω) FOR THE LED.

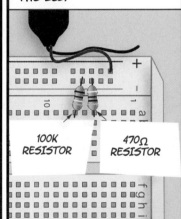

100K RESISTOR

470Ω RESISTOR

ADD THE LED. CONNECT THE ANODE (LONG LEAD) TO THE 470Ω RESISTOR.

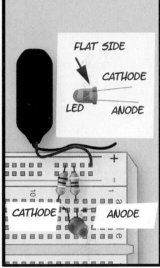

FLAT SIDE

CATHODE

LED

ANODE

CATHODE

ANODE

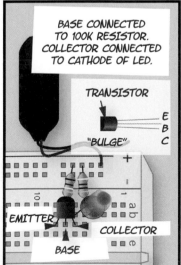

BASE CONNECTED TO 100K RESISTOR. COLLECTOR CONNECTED TO CATHODE OF LED.

TRANSISTOR

"BULGE"

E
B
C

EMITTER

COLLECTOR

BASE

TO TURN THE LED ON AND OFF, THIS CIRCUIT USES AN NPN TRANSISTOR. MAKE SURE THE "BULGE" IS FACING THE RIGHT WAY.

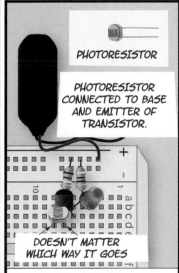

PHOTORESISTOR

PHOTORESISTOR CONNECTED TO BASE AND EMITTER OF TRANSISTOR.

DOESN'T MATTER WHICH WAY IT GOES

THE PHOTORESISTOR SENSES WHETHER IT'S LIGHT OR DARK. CONNECT IT BETWEEN THE TRANSISTOR'S BASE AND EMITTER.

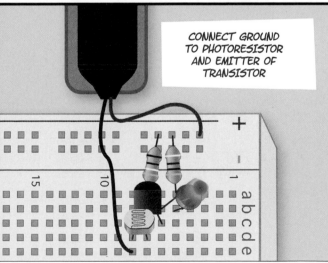

CONNECT GROUND TO PHOTORESISTOR AND EMITTER OF TRANSISTOR

FINALLY, CONNECT THE BLACK LEAD OF THE BATTERY CLIP (GROUND). ATTACH A BATTERY TO THE CLIP. WHAT HAPPENS WHEN YOU COVER THE PHOTORESISTOR WITH YOUR THUMB? WHAT HAPPENS WHEN YOU SHINE A LIGHT ON IT? HOW COULD YOU REARRANGE THESE COMPONENTS TO MAKE A "LIGHT DETECTOR"?

GoPro Hero 4-Series Cameras

$ varies : gopro.com

The newest breed of GoPro cameras have every feature ever asked of them, minus the one that always made them special.

Every camera is a compromise. Anyone who wants to capture the world around them wants a camera that is easy to use, produces stunning images, and is light on the pocketbook. GoPro launched their cameras several years ago to fill the unique niche of go-anywhere cameras that were cheap to buy, easy to use, and offered a great image. Their Achilles' heel was that their image quality wasn't quite good enough to meet the expectations of high-end users.

While last year's Hero 3+ brought those discussions down to a murmur, the Hero 4 Silver and Black cameras finally silence the argument. Shooting up to 4k video at 30 fps (Black edition only), the new cameras offer professional-level quality, and finally give users some control over exposure and color balance. The new nighttime photography settings provide incredible long-exposure and time-lapse options.

But where these cameras compromise is their cost. It would be a stretch to call the previous cameras disposable, but they were at least expendable. At $400 and $500 for the Hero 4 Silver and Black editions respectively, you're likely to do a lot of hemming and hawing before you hang your Hero 4 out in the wind — even though that's where GoPro cameras do their best work. And while the new base model GoPro Hero camera still hits that expendable price point, it's missing so many of the other features that make the more expensive cameras so excellent. —*Tyler Winegarner*

Hep Svadja

SHAVIV DEBURRING TOOL SET

$20 : vargus.com/shaviv

Don't worry if you have never heard of deburring tools before, as they tend to be industrial-type tools that you won't find at the local big-box, home-improvement store. That said, the first time you use a deburring tool, you will curse yourself for all those times you removed sharp edges and burs with a file — or worse, with sandpaper.

This Shaviv Mango II assortment gives you all you need to get started: a swivel handle, a couple of general-purpose blades that cover most common shapes and materials, and an extending blade holder for reaching into tighter spaces.

—*Stuart Deutsch*

HOWARD LEIGHT BY HONEYWELL SYNC STEREO EARMUFFS

$36 : howardleight.com/ear-muffs/sync

As a busy dad with a short commute, I seldom get to just rock out anymore — really escape into "the metal." The only exception is when I'm doing chores with obnoxiously loud power tools. I've used Bluetooth headphones and loved the convenience, but their volume couldn't compete with the noise of a weed whacker or other loud tools. My ears were doubly blasted, my enjoyment of Van Halen halved.

Now I'm loving the Sync earmuff — an industrial-grade hearing protector styled like over-ear DJ headphones, with high-quality audio built in (and a removable audio cord for when you just want earmuffs). They automatically cap the audio volume at a safe 82dB, but that's all you'll need, as the 25dB noise reduction rating means you'll be blissfully untroubled by the racket of your angle grinder, rototiller, or toddler's tantrums while attempting to get things done around the house or workshop.

—*Keith Hammond*

BONDHUS BALL HEX ALLEN WRENCHES

$26 : bondhus.com

Ball hex drivers are the way to go for robotics projects and anywhere else you might need to access hex fasteners. This Bondhus set comes with separate inch and metric sets with GoldGuard and BriteGuard finishes that make the L-wrenches easy to clean and identify.

The straight hex on the short end is best for higher torque applications, and the ball hex tip allows for quicker spinning and easier access. Instead of having to enter a hex-head screw straight on, ball hex tips can fit screws at a slight angle.

These Bondhus drivers are strong, durable, and very inexpensive for USA-made tools.

—*SD*

BLACKFIRE CLAMPLIGHTS

$28+ : blackfire-usa.com

The Blackfire Clamplight is a handy hands-free LED flashlight that's built around a powerful 1-watt Cree LED that outputs up to 100 lumens for 26 hours on a single set of 3 AAA batteries. It has decent performance stats, but where the Blackfire really excels is in the packaging. The flashlight's body is a spring-loaded clamp that allows it to grab onto a range of objects, such as a pipe or the edge of a table. The head pivots around on two axes to provide illumination right where you need it. If there is nothing for the flashlight to clamp onto, flip the clamp pads down, and the mini stand allows the light to sit unsupported on a floor or tabletop.

—*Eric Weinhoffer*

Stuart Deutsch

MAXXPACKS CUSTOM BATTERY PACKS

$ varies : maxxpacks.com

For some projects, a AA or AAA battery holder and a few alkaline or rechargeable batteries will do. But for projects that involve motors or high-drain components, finding a suitable battery pack isn't always easy.

I needed two rectangular block-shaped 9.6V NiMH battery packs for a recent robotics project, and could not find any off-the-shelf products that were perfectly fitting. I ultimately learned about MaxxPacks and found a suitable pack in their online store. Not only that, they were able to build it with the connector I needed too.

Buying the battery packs from MaxxPacks was a great experience, and I'm satisfied with the quality and quick service. The control freak in me loves that I knew everything about the chosen battery packs before I ordered, including discharge rate, capacity, and even the brand of cells the packs were built with.

—*SD*

Gunther Kirsch

VAMPLIERS MINI

$30 : vampiretools.com

Like the other Vampire Tools I've used, the Vampliers Mini are rugged, made from high-quality materials, and simply a joy to use. Although smaller and slightly less flexible in their use than the "original" Vampliers, the Minis are made of equally high-quality, ESD-safe handle materials and fabricated flawlessly, with vertical serrations for gripping screws. Unlike the original, there's a key-ring hole on the handles for transport, and the lack of a spring makes them easier to store in their closed state.

I'd probably still recommend the standard Vampliers to anyone who doesn't spend their whole day working on smaller electronic projects, but these pliers are fantastic for general plying tasks and screw extraction, making them a great addition to anyone's toolbox.

—*EW*

NEW MAKER TECH

MICROVIEW

$40 : sparkfun.com

It can be tough to know what your Arduino is thinking sometimes. That's why the folks at Geek Ammo created MicroView, a chip-sized Arduino-compatible module with a built-in 64x48 pixel OLED display on top. It has 12 digital I/O pins, three of which are capable of PWM, six analog inputs, and can be powered with 3.3 to 16 volts. Since the development board is in a DIP package, it can be easily pushed into a breadboard for prototyping.

The display is controlled with their Arduino library, which makes it easy to draw text, sprites, graphs, and gauges on the screen for interactive menus, readouts, or a quick look at what's going on inside that chip. —*Matt Richardson*

POLOLU DRV8835 DUAL MOTOR DRIVER SHIELD FOR ARDUINO

$7 : pololu.com

There are numerous Arduino-compatible motor shields that allow for easy control of motors, servos, and stepper motors, but most are one-size-fits-all designs that are meant to satisfy many different needs. Pololu's new DRV8835 dual motor driver shield is a simpler, smaller, and very inexpensive shield that can be used to control one or two small brushed DC motors.

The shield can deliver 1.2 A continuously (1.5 A peak) along both channels, or 2.4 A (3 A peak) when both channels are connected together in parallel. It features six screw-down terminals — two are for an external 1.5 V to 11 V power supply, and four for the pair of motor control channels. —*SD*

NAVIO AUTOPILOT SHIELD

$150 : emlid.com

There are a number of sensors, functions, and features that you will find common to many autonomous robotics projects: GPS navigation, accelerometer and gyroscope 3D-position and orientation sensors, motor controllers, analog inputs, and a wireless receiver to name a few. The Navio Autopilot Shield from EMLID packs all these features and more into a small Raspberry Pi daughter board. Navio was designed as an experimental hardware platform for the Linux version of the ArduPilot APM Autopilot, an open-source software suite for autonomous robotics.

The Navio can help you build the controller for your plane, car, boat, or drone. And since it works on top of the Raspberry Pi, you can easily use other software libraries to extend the functionality of your autonomous robotic project, such as with Wi-Fi, a webcam, GSM, and more. —*MR*

ENOCEAN SENSOR KIT AND ENOCEAN PI

$70 (kit), $26 (add-on board) : element14.com

When you work with wireless sensors, one of the challenges is figuring out how you power each sensor unit. But with the EnOcean Sensor Kit and the EnOcean Pi add-on board, the only thing you'll need to power directly is the Raspberry Pi. Each wireless sensor module harvests its own electricity, either from ambient light or kinetic energy. The sensor kit from element14 includes a reed switch, a temperature sensor, and a pushbutton switch.

The EnOcean Pi receives the wireless signals from the sensors and passes them to a Raspberry Pi via serial. EnOcean provides a guide to reading the data in FHEM, an open-source server for home automation. For experienced coders, you can have these sensors working with your projects no matter what language they're written in. —*MR*

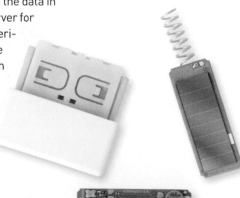

NEW MAKER TECH

RASPBERRY PI MODEL A+

$20 : raspberrypi.org

Not long after upgrading the Raspberry Pi Model B to the Model B+, the foundation behind the popular single board computer has now brought a lot of the same improvements to their low-end line with the new Model A+. The A+ sports the newly expanded 40 pin breakout, it works with microSD cards, and the new board now has rounded corners with mounting holes. The smaller PCB footprint and slim profile make the A+ a great option for projects where space is tight.

As far as performance goes, the A+ has similar specs as the Model A. It has the same Broadcom BCM2835 system on a chip (SoC) with 256 megabytes of RAM. Priced at just $20, the refreshed Model A+ keeps Raspberry Pi in the lead in terms of affordability in the landscape of Linux-based boards. —MR

GIZMO 2 SINGLE BOARD COMPUTER

$199 : gizmosphere.org

Packing a lot of power in just 4 inches squared, the Gizmo 2 from Gizmosphere is an open-source single board computer aimed at professional embedded programmers and advanced makers. It has a 1 GHz AMD GX-210HA x86 dual-core processor with a 300 MHz GPU and 1 GB of RAM, which puts the Gizmo 2 on the high end of maker-friendly single board computers.

Not only is it powerful, but the Gizmo 2 also supports many connectivity interfaces. With an HD audio input and output, mSATA, mini PCIe, microSD, gigabit Ethernet, and USB port, there's not much this small board can't do. And since the board is open source, the schematics, PCB layout files, and bill-of-materials are available if you'd like to use the design to build your own version. —MR

PRINT

FIVE HUNDRED AND SEVEN MECHANICAL MOVEMENTS POSTER SET

By Henry T. Brown; Poster by Liz Rettger
$35–$60 : etsy.com/shop/RettgerGalactic

Henry T. Brown's illustrated book *507 Mechanical Movements* is more than a field guide to a menagerie of mechanisms. It's kind of a Rosetta stone for machinery, showing how you can translate virtually any sort of motion to any other. Worm drives, planetary gears, rack and pinion, clutch boxes, eccentric cams, valves and governors, on and on, with succinct descriptions of how each one works. Published almost 150 years ago, it's a window into the Industrial Revolution that's still a touchstone for inventors and machinists today. It's free on Google Books, but for those who'd like to drink in all the visual richness at once, Liz Rettger has laid out all 507 illustrations on two (or three) wall posters that would grace any maker's space. —KH

DIY RC AIRPLANES FROM SCRATCH: THE BROOKLYN AERODROME BIBLE FOR HACKING THE SKIES

By Breck Baldwin $16 : Tab Books

If you have been to World Maker Faire in New York, you might have seen Brooklyn Aerodrome's cheerful triangular airplanes soaring and looping around in the sky. BA sells airplane kits made out of sheets of foam insulation, off-the-shelf motors, and standard transmitters and receivers. The kits are a very economical and kid-friendly entry into learning how to build your own RC airplane. The Brooklyn Aerodrome Bible is a summary of what the group learned in producing their airplane design. It walks you through the kit, offering bug fixes and hardware alternatives, and shows photos of past plane designs. It's a fantastic guidebook for building inexpensive RC planes.

— John Baichtal

THE MAKER'S MANUAL

$34.99 : By Paolo Aliverti and Andrea Maietta

The Maker's Manual is a practical and comprehensive guide to becoming a hero of the new industrial revolution. It features hundreds of color images, techniques to transform your ideas into physical projects, and must-have skills like electronics prototyping, 3D printing, and programming. This book's clear, precise explanations will help you unleash your creativity, make successful projects, and work toward a sustainable maker business.

MAKE: A RASPBERRY PI-CONTROLLED ROBOT

$19.99 : By Wolfram Donat

Make a Raspberry-Pi Controlled Robot teaches you how to build a capable and upgradeable personal robot for around $100. You'll learn how to control servos, respond to sensor input, and know where your bot is using GPS. You'll also learn many ways to connect to your robot and send it instructions, from an SSH connection to sending text messages from your phone.

21st CENTURY ROBOT

$24.99 : By Brian David Johnson

When companies develop a new technology, do they ask how it might affect the people who will actually use it? That, more or less, sums up Brian David Johnson's duties as Intel's futurist-in-residence. In this fascinating book, Johnson provides a collection of science-fiction prototyping stories that attempt to answer the question.

GETTING STARTED WITH ARDUINO — 3RD EDITION

$19.99 : By Massimo Banzi and Michael Shiloh

Arduino is the hot open-source prototyping platform for artists, hobbyists, students, and anyone who wants to create interactive physical environments. *Getting Started with Arduino* is co-authored by Arduino co-founder Massimo Banzi, and incorporates his experience in teaching, using, and creating Arduino. New in this edition is a chapter on Leonardo and an irrigation project for plants and gardens.

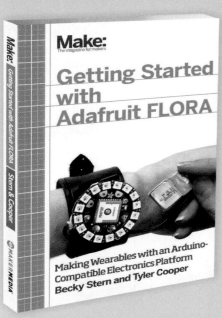

GETTING STARTED WITH ADAFRUIT FLORA

$16.99 : By Becky Stern and Tyler Cooper

This book introduces readers to building wearable electronics projects using Adafruit's tiny Flora board: at 4.4 grams, only 1.75 inches in diameter, and featuring Arduino compatibility, it's the most beginner-friendly way to create wearable projects (see pages 35 and 44 for more information). This book shows you how to plan your wearable circuits, sew with electronics, and write programs that run on the Flora to control the electronics. The Flora family includes an assortment of sensors, as well as RGB LEDs that let you add lighting to your wearable projects.

ENCYCLOPEDIA OF ELECTRONIC COMPONENTS VOL.2

$29.99 : By Charles Platt

Want to know how to use an electronic component? This second book of a three-volume set includes key information on electronics parts for your projects — complete with photographs, schematics, and diagrams. You'll learn what each one does, how it works, why it's useful, and what variants exist. Volume 2 covers signal processing, including LEDs, LCDs, audio, thyristors, digital logic, and amplification.

GETTING STARTED WITH CNC

$16.99 : By Edward Ford

Getting Started with CNC is the definitive introduction to working with affordable desktop and benchtop CNCs, written by the creator of the popular open-hardware CNC, the Shapeoko. Although inexpensive 3D printers can make great things with plastic, a CNC can carve highly durable pieces out of a block of aluminum, wood, and other materials. This book covers the fundamentals of designing for — and working with — affordable CNCs.

CUBE 3

WRITTEN BY JOHN ABELLA

Has the 3D printer as an appliance finally arrived?

Cube 3 | cubify.com

- **Price as Tested** $999
- **Build Volume** 734×523×401mm
- **Bed Style** Aluminum plate with bonded plastic top layer
- **Temperature Control?** No
- **Materials** Proprietary ABS or PLA
- **Print Untethered?** via wi-fi or USB thumb drive
- **Onboard controls?** Color touch screen
- **Software** Cubify
- **Slicer** Cubify
- **OS** Windows / Mac / iOS / Android
- **Open Software?** No
- **Open Hardware?** No

PRINT SCORE: 29

● **Accuracy**	1 2 3 **4** 5
● **Backlash**	1 2 3 4 **5**
● **Bridging**	1 2 3 **4** 5
● **Overhangs**	1 2 3 **4** 5
● **Fine Features**	1 2 3 4 **5**
● **Surface Curved**	1 2 **3** 4 5
● **Surface General**	1 **2** 3 4 5
● **Tolerance**	1 2 3 4 5
● **XY Resonance**	FAIL **PASS (2)**
● **Z Resonance**	**FAIL** PASS (0)

PRO TIPS

● The intended surface preparation for prints is to put down a layer of the included CubeGlue prior to each print, and to wash that glue off between prints as it interferes with the optical bed-leveling routine. I was able to reuse the glue between a number of prints by skipping the leveling. With the glue I didn't have any curling problems with large, lengthy prints.

WHY TO BUY

3D Systems third-generation Cube printer is a real contender for the home-printer-as-appliance crown. It knows when you install ABS or PLA cartridges, what color they are, and adjusts accordingly. It tells you how much filament you've used, so it'll warn if you try to print something too large for what you have left.

How'd it print?

JOHN ABELLA is an obsessive hobbyist and 3D printing enthusiast. He teaches 3D printer assembly workshops with BotBuilder.net

THE CUBE 3'S FEATURE CHECKLIST MAKES IT STAND OUT IN A SEA OF NEW MACHINES. Color touchscreen, printing over wi-fi or USB stick, printing from iOS or Android devices, autoleveling, auto-z-height calibration, and dual-color printing are all standard features.

NEW NOZZLE WITH EVERY CARTRIDGE

The Cube 3 only works with chipped, proprietary filament cartridges, which cost more and hold less plastic than spooled filaments, but each includes a completely new nozzle, minimizing the chance for jams.

RELIABLE DUAL EXTRUSION

There is no other printer in this price range that offers dual-color capability with the ease and reliability I experienced with the Cube 3. I was able to take normal STL files, individually color parts of them, and have them print in dual colors without complicated alignment or tedious STL file merging. Printing in two colors is slow, but it's hard to argue with the results, which were mostly good.

NICE DESIGN, BUT NOISY AND SLOW

I'd estimate that it's one of the noisiest machines we've tested — not something you want running while you watch TV. The case resonates, and it's common for the various motors to attempt to move past their limits during initial (nonprinting) movements. Prints take a long time, but have a high success rate — a balance that makes sense for entry-level users.

The industrial design for the price is miles ahead of the competition. Integrated spool holders, optical bed leveling, integrated lighting, and the use of linear rails in a package this affordable has raised the bar. Special care was taken to ensure that it's almost impossible for curious youngsters to get pinched or burned by the machine.

NO TWEAKING

The printer, software, and filament are all tightly controlled to simplify the experience. On the flip side, the quality you get won't ever get better due to your own tweaking. The end user never sees actual G-code, and variables like temperature and speed are kept under wraps.

CONCLUSION

This is not a machine for people who want to tweak settings and chase the perfect print. The Cube 3 is for someone who wants to print, has no time to waste, and is willing to pay more for plastic for a toolchain that just works. ●

CUBEPRO

WRITTEN BY MATT STULTZ

This large-build, high-end machine is not your average home printer

CubePro | cubify.com

- **Price as Tested** $4,399
- **Build Volume** 200.4×230×270.4mm
- **Bed Style** Unheated proprietary material bed, heated build chamber
- **Temperature Control?** No
- **Materials** Proprietary ABS, PLA, Nylon cartridges
- **Print Untethered?** Via wi-fi or USB stick
- **Onboard controls?** LCD touchscreen
- **Software** CubePro software
- **Slicer** CubePro software
- **OS** Mac, Windows
- **Open Software?** No
- **Open Hardware?** No

PRINT SCORE: 28

● Accuracy	1 2 3 4 **5**	
● Backlash	1 2 **3** 4 5	
● Bridging	**1** 2 3 4 5	
● Overhangs	1 **2** 3 4 5	
● Fine Features	1 2 **3** 4 5	
● Surface Curved	1 2 3 **4** 5	
● Surface General	1 **2** 3 4 5	
● Tolerance	1 2 **3** 4 5	
● XY Resonance	FAIL **PASS (2)**	
● Z Resonance	FAIL **PASS (2)**	

PRO TIPS

- Let your glue dry a little after applying it to keep your prints from sliding off.
- You must unload all PLA before you perform an ABS print, so choose your materials before loading the machine.

WHY TO BUY

This is a large-build-volume, professionally designed machine that really targets the gap between printers designed for the home user and those designed for the professional markets. Accessible via wi-fi and can either connect to an existing network or create an ad hoc network.

FROM THE MOMENT YOUR 90-POUND CUBEPRO ARRIVES ON A WOOD PALLET VIA A FREIGHT TRUCK, YOU'LL KNOW THIS IS NOT YOUR AVERAGE HOME PRINTER. 3D Systems has created a machine for both the serious printing aficionado and companies who are looking for a more robust and professionally built printer.

MULTIPLE EXTRUDERS, MULTIPLE CHIPPED MATERIALS

Single, double (Duo), or triple (Trio) extruder models are available, but each additional extruder diminishes the build area (single extruder build width of 285mm, triple 200mm). These multiple extruders deliver multiple materials. 3D Systems ships cartridges of ABS, PLA, and nylon for the CubePro. The embedded microchips tell the machine what color and material is loaded and how much is left, but also prevent the use of nonofficial filaments.

HEATED BUILD CHAMBER

Desktop 3D printers use heated build platforms to prevent ABS prints from curling and warping off the build surfaces. However, exposed print areas are susceptible to drafts causing uneven cooling and cracking. The CubePro uses the same heated, enclosed design employed by high-end professional machines to prevent this.

SOFTWARE SHORTCOMINGS

Although the CubePro software is easy to use, it has some severe limitations. Print options are constrained to three levels of print quality and infill. Provided estimates are often far off from the actual print times. The greatest discrepancy I encountered was an estimate of 45 minutes that resulted in a 2 hour and 30 minute print.

With PLA prints, high print temperatures (uncontrollable by the user) cause anything close to an overhang to droop. Our bridging test failed on all levels. Rounded- or sloped-top prints looked very nice, but prints with large, flat surface tops showed signs of over extruding or bed-height issues. However, I've been a CubePro beta tester and seen the print quality continually improve over time through software updates.

CONCLUSION

The CubePro would sit nicely in an office beside the company's large copy/scan/print appliance, where changing the filament cartridges would be akin to changing the toner cartridge. However, most home users will have a hard time finding space for it or needing quite as much machine as the CubePro provides. ◐

How'd it print?

MATT STULTZ is a community organizer and founder of both 3D Printing Providence and HackPittsburgh. He's a professional software developer, which helps fuel his passion for being a maker. 3DPPVD.org

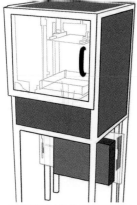
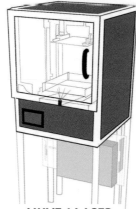

"ACME Catalog, Page 115: Battle Cat"

Written by James Burke

"AH, AN EXCELLENT CHOICE SIR. This is one of our finest quality products, and a brand-new addition for our ACME Spring Catalog. Are there any specific questions you have? Well of course: It's the result of decades of research from armor specialist Jeff de Boer.

I assure you he knows what he's doing. His work is currently under review for authenticity by the Royal Armouries in Leeds. They certainly don't accept anything less than perfection. He was also born into a tinsmithing family. Yes, he did graduate from the Alberta College of Art and Design with a major in jewelry, but he later switched to feline defense armaments full time. There is quite a variety of meticulously crafted armor, ranging from medieval period designs to feudal samurai. If you desire a custom piece, you're looking at about 120 to 300 hours of work.

I'm certain this will catch birds off guard. (Though be aware, de Boer does produce protective gear for mice.) We have a few in stock and can rush an order; if you place it on the phone with me right now you'll get our trademark instantaneous courier service. You can always count on ACME to deliver the zenith of any industry, be it needlessly specific roadrunner traps or inexplicably volatile explosives. All right? Yes. You can give me your card number as soon as you're ready. Do you have a last name, Sylvester?"

Jeff de Boer